T0132816

WATER IN TEXAS

TEXAS NATURAL HISTORY GUIDES ™

WATER IN TEXAS

AN INTRODUCTION

ANDREW SANSOM

With Emily R. Armitano and Tom Wassenich

UNIVERSITY OF TEXAS PRESS
Austin

This publication received support from the Peter T. Flawn Series in Natural Resource Management and Conservation and the RGK Foundation.

Requests for permission to reproduce material from this work should be sent to:

Permissions
University of Texas Press
P.O. Box 7819
Austin, TX 78713-7819
http://utpress.utexas.edu/index.php/rp-form

♾ The paper used in this book meets the minimum requirements of ANSI/ NISO Z39.48-1992 (R1997) (Permanence of Paper).

LIBRARY OF CONGRESS CATALOGING-IN-PUBLICATION DATA

Sansom, Andrew.
   Water in Texas : an introduction / Andrew Sansom ; with Emily R. Armitano and Tom Wassenich.
      p. cm. — (Texas natural history guides)
   Includes bibliographical references and index.
   ISBN 978-0-292-71809-8 (pbk. : alk. paper)
   1. Water—Pollution—Texas. 2. Water-supply—Texas. I. Armitano, Emily R. II. Wassenich, Tom. III. Title.
   TD224.T4S26 2008
   363.6'109764—dc22   2007048816

For Tim and
Karen Hixon,
whose devotion to
conservation in Texas
is unsurpassed

CONTENTS

FOREWORD

Denise M. Trauth, President, Texas State University

When I set foot on the grounds of Texas State University for the first time, one of the things that impressed me the most was the headwaters of the San Marcos River. It was hot, and I remember how refreshing the water looked and how beautiful it was. All kinds of people were enjoying the river and its banks. I quickly learned how significant it is in the life of the university and the community, how unique it is, and that it must not be taken for granted.

Surveys of our alumni indicate that the river is what they miss the most after graduating and leaving San Marcos. Its presence and stewardship are central to our identity, and it inspires a core value in our education, research, and service. We are the water university, and the stewardship of water resources in general is central to our academic identity.

Thus it is fitting that this new work has been created by a member of our family. Andrew Sansom, research professor of geography at Texas State and director of our River Systems Institute, has compiled the first comprehensive guide to water in Texas, including science, natural and cultural history, geography, and public policy. Sansom's efforts on campus include support of a renowned Ph.D. program in aquatic resources, which prepares leaders to bring together various disciplines in the research, management, policy making, and education surrounding sustainable water issues, not only in Texas, but also worldwide.

Texas is at a crossroads. The choices we make now will determine whether we will be able to meet our water needs to enhance our economy and quality of life and protect our natural resources. *Water in Texas* is a guide to help us find the way.

ACKNOWLEDGMENTS

I first met Tim and Karen Hixon twenty-five years ago when we ran the Snake River in Idaho through the Birds of Prey Natural Area. We shared an extraordinary experience on the river and have remained close friends and allies in conservation ever since. Tim and Karen are among the most unassuming people I know, yet their combined contributions to our movement in time, leadership, philanthropy, and just plain encouragement reach well beyond Texas to all parts of the world. They are the second couple ever to have served on the prestigious Texas Parks and Wildlife Commission, and without their underwriting this book would not have been possible. I am grateful to them for that, of course, and for all they have done for the outdoors, but I am most grateful that my kids got to know two such fine human beings.

Many other people contributed to this book in many ways. It is truly a collaboration. My longtime colleague, Emily Armitano, served as a kind of producer, taskmaster, and editor; and Tom Wassenich headed up the research.

Our team was fortunate to have funding from the Meadows Foundation and to have had the benefit of a number of expert reviewers, including Rich Earl, Warren Pulich, Joni Charles, Norman Johns, David Bradsby, Cindy Loeffler, Jason Pinchback, and Myron Hess. The graphics were skillfully executed by two of my students, Eduardo Valdez and Jessica Spangler. Two other talented students, Mary Waters and Lauren Bilbe, fact-checked the text. Chad Norris, Jennifer Ellis, Tyson Broad, Fran Sage, and Susan Romanella helped whenever we called on them.

I am deeply grateful to Wyman Meinzer, the Texas Parks and Wildlife Department, Blake Matejowski, Gregg Eckhart, and several others for the many wonderful images that help to tell this story so well.

Finally, working with the University of Texas Press is a pleasure, thanks to Bill Bishel and his team, including Megan Giller. This is my second project with them; I truly hope it will not be the last. My heartfelt thanks to everyone who made this book possible.

WATER IN TEXAS

Cypress trees line the Sabinal River.
Photo by Wyman Meinzer.

1. INTRODUCTION

Living with a Limited Resource

No natural resource has greater significance for the future of Texas than water. For nearly fourteen years, twelve as executive director, I had the privilege of working at the Texas Parks and Wildlife Department. During that time, I was able to see more of the richness of Texas' cultural and natural history than most people see in a lifetime, and I came to have a profound respect for water's role as the limiting factor for all of life and the principal determinant of economic progress.

In Texas we have been very successful in managing our natural resources over the past century or so. David Schmidly's work, *Texas Natural History: A Century of Change*, has shown that in most cases the landscape of our state is in much better condition than it was before 1900. Barely fifty years after settlement began in earnest with the arrival of Anglo colonists and sodbusters, most of the native grasslands had been plowed under or overgrazed and the vast virgin woodlands of the Piney Woods deforested. Great quantities of soil had washed off the land, especially in the Hill Country and the Rolling Plains.

Today, thanks to the efforts of government agencies such as the Agricultural Extension Service and the Soil Conservation Service (now the Natural Resource Conservation Service) and the good stewardship of private landowners, who own most of the real estate in Texas, the condition of the landscape has improved. Water quality has improved as well. The establishment of pollution control legislation has had a positive impact on many of our rivers and streams, which until the late 1960s were often contaminated with poorly treated or untreated industrial and municipal waste.

In addition, our fish and wildlife populations are in better condition than they were at the beginning of the twentieth century. It is hard to believe, but at one time whitetail deer were largely extinct in parts of the state; today in some areas their population has increased to the extent that they present a serious ecological problem. Thanks to sound wildlife management, each year we harvest more wild turkeys than existed in the entire state before World War II. We have stocked more than a billion fish in our waters and substantially eliminated or fundamentally limited commercial harvest of marine and freshwater species.

Texas is the number one hunting and the number three fishing state in the nation. It is among the best destinations for bird-

Hunters at sunrise. Photo courtesy of Texas Parks and Wildlife Department.

watchers in the world. It possesses a system of state parks and wildlife management areas that is the envy of other states, and it is home to some of the nation's most important national parks and national wildlife refuges. Thus we have reason to feel good about the natural condition of our state. But we must be prepared to confront very serious challenges both on the land and in the water.

With respect to the landscape of Texas, the most important fact and insight is that the state is almost entirely in private ownership. Experts argue about the exact percentages, but it is indisputable that between 94 and 97 percent of the state is privately held. The implications for the environment, water resource management, wildlife, outdoor recreation, and open space are profound. In addition, Texas has become one of the most urbanized states in the country and, as a result, loses large tracts of rural and agricultural land to development each year. Often, as traditional landowners pass away, their heirs are left with as much tax burden as land. All these factors contribute to an inexorable process of land fragmentation, which is the single greatest terrestrial environmental problem we face.

As the size of tracts of land in Texas continues to diminish, wildlife habitat disappears and open space is lost, along with much of the outdoor recreation opportunities we enjoy. Perhaps most noteworthy for the future of the state, the function of our watersheds is irrevocably impaired. In fact, the issues associated with ensuring sufficient clean water for both economic growth and the environment is the most significant and urgent environmental concern facing Texas in the next generation.

Another significant insight is that most of the available water is in eastern Texas, while most of the current and expected economic growth is in the west where water is scarce. Over the years there has been much talk of moving large amounts of water from the eastern rivers westward to thirsty farms and cities, but regional competition, environmental objections, and substantial legal impediments have kept this from happening on a major scale.

Historically, we have depended largely on surface water for

human uses, including agriculture, industry, municipal consumption, and recreation. Surface water occurs naturally in the rivers and is stored in 196 major reservoirs—one of the most extensive series of impoundments in the United States. According to the 2007 State Water Plan, a major reservoir is defined as one that has at least 5,000 acre-feet of storage capacity at its normal operating level.

By law the water in the rivers is considered the property of the people of Texas. Today most of this water has already been spoken for through the granting of rights to it by the state of Texas. In fact, some of our rivers are actually overappropriated; that is, if all the water permitted for use from them were withdrawn they would dry up. Anyone who doubts this assertion may simply recall photographs from the beginning of the twenty-first century showing that the Rio Grande no longer reached the Gulf of Mexico. Furthermore, about 85 percent of surface water has been permitted for agricultural use, rendering its availability for purposes such as urban development very difficult.

Compounding this problem is the fact that the state has not constructed a new reservoir in about fifteen years. The most recent is the O. H. Ivie Reservoir on the Concho River, dedicated in 1990. There just are not many sites left in Texas where reservoirs can be built, and those that exist often have important associated environmental resources, including much of the state's remaining bottomland hardwood forests and other significant wildlife habitat. The latest state water plan (water plans are issued every five years) envisions fourteen new major impoundments, and there are already concerns about many of these proposed projects. Finally, private landowners in Texas have been able to take advantage of the current state water planning process to discourage the taking of their property for reservoir construction. In sum, all these issues make the process of approval for reservoir construction a challenge that can take many years to complete.

Another important factor constraining the use of surface water is that the state has provided very little protection histori-

The Rio Grande failed to reach the Gulf of Mexico in 2001. Photo courtesy of Texas Parks and Wildlife Department.

cally for what are called environmental flows, the amounts of water necessary to sustain aquatic life in the rivers and bays and the estuaries into which they empty. The Texas legislature did not officially recognize that protecting the aquatic environment was a beneficial use of water until 1985, when fairly modest provisions were included in water rights permits for the first time to protect environmental flows. By that time, unfortunately, the vast majority of Texas' surface water had already been permitted for use. This is a problem for both developers of water resources and for environmental interests, who joined forces in 2007 to encourage the state legislature to lay the groundwork for protecting environmental flows. From an environmental standpoint, if we are not able to sustain the flow of freshwater into the bays and estuaries, their biological productivity will decline substantially. These areas provide not only the best coastal sport fishery in the country but also billions of dollars of annual economic benefit to the state through waterfowl hunting, bird-watching, and recreational and commercial fishing.

Largely as a result of these daunting challenges to the limited surface water supply, Texas is increasingly looking to groundwater as its principal source of water. Groundwater use in Texas is not a new concept. San Antonio, for example, has historically been 100 percent dependent on groundwater from the Edwards Aquifer for both industrial and municipal use. A primary reason for this is that for nearly one hundred years Texas did not regulate groundwater use.

A Texas Supreme Court decision in the early twentieth century declared that groundwater was too "mysterious and occult" to understand and thus to regulate. Since then the rule for groundwater use in Texas has been the right of capture, which means that anyone who owns land above a subterranean water reservoir can pump an unlimited amount of water for any purpose. This total lack of regulation of groundwater is in stark contrast to the heavy regulation of surface water and is the primary reason that entrepreneurs in various parts of the state, including the Panhandle, the Trans-Pecos, and Central Texas, are feverishly attempting to secure and market very large volumes of water from aquifers.

A complicating factor is that Texas' nine major and twenty-one minor subsurface aquifers are very different geologically. Some aquifers recharge themselves fairly regularly; others were charged as long ago as the Ice Age and are not sustainable. In recent years the Texas legislature has enabled the establishment of local groundwater management districts to begin bringing some semblance of order to groundwater use in the state. However, many of these districts are organized on county lines rather than on the natural boundaries of the aquifers, are very poorly funded, and lack either the fundamental science or expertise to do their jobs.

Most scientists concur that there are substantial groundwater reserves available that directly affect, and even sustain, our surface waters. But without laws and policies linking ground- and surface water, sustainable management of these resources will be ineffective.

Traditionally, it has taken a bad scare to get politicians in Texas to address the state's water problems. Much of the existing water infrastructure and the planning process on which the future of the state depends are the result of the drought of the 1950s, the so-called drought of record. In the 1990s Texas experienced a drought that in many respects was as serious as the previous crisis. But by the end of the twentieth century most Texans had migrated to urban areas where the effects of such a drought were not so obviously experienced when everyone lived on farms or ranches or in small towns.

Nevertheless, the drought of the 1990s sparked a new spate of water-related laws that provide the context for addressing Texas' future water needs in the new century. Senate Bill 1, passed in 1997, is considered a landmark piece of legislation; its centerpiece is the creation of a bottom-up planning process that involves local interests and stakeholders through regional planning committees, replacing the old centralized planning system that came into being after the drought of record. Unfortunately, when these regional planning groups were established many of our river basins and watersheds were divided, making system-wide planning very difficult. In addition, some of the groups have considered environmental issues important while others have ignored them altogether, rendering their conclusions ineffective at best and destructive at worst.

As Texas has continued to urbanize, another disturbing trend in recent years is the increasing lack of consideration for issues of concern to rural areas and small towns. Dallas, for example, imposes unwanted reservoirs on East Texans, and San Antonio continues its reliance on the Edwards Aquifer, which threatens spring flows in San Marcos and New Braunfels.

The planning system put in place through Senate Bill 1 demonstrated that groundwater is going to have to be a much bigger part of the equation, and it was the basis for Senate Bill 2. This

Dry creek bed in West Texas. Photo courtesy of Texas Parks and Wildlife Department.

bill, enacted in 2001, enabled the creation of groundwater management districts throughout the state, thus amending the rule of capture for the first time in a century. Finally, in 2007, the Texas legislature took another bold step and established, for the first time, a process for protection of environmental flows in the state's rivers and streams with Senate Bill 3.

Thanks to the widespread public participation mandated by

the new laws, to the growth in population, and to the obvious fact that some of our most important sources of water are increasingly limited, water remains in the forefront of public policy in Texas. Unfortunately, we still seem to require a looming catastrophe to spur action.

One of the most promising ways to extend our supplies of water is through increased efficiency and conservation in both agricultural and municipal use. The city of San Antonio has made major strides in reducing its consumption of water, while consumption in other Texas communities has continued to grow. Certain agricultural sectors, in particular in the High Plains, have dramatically improved the efficiency of irrigation practices, while the production of other farm commodities, including citrus, is way behind. Many cities and water authorities are looking closely at reuse of treated wastewater, which, while intuitively logical, could cause problems for communities and interests downstream that depend on return flows.

Finally, looming over all of this is the fact that all modern water planning in Texas for the past fifty years has been based on the notion that the drought of the 1950s is as bad as it is going to get. Today, with the widespread consensus that the climate is changing as a result of both natural and human-caused phenomena, such thinking is not acceptable. Examination of fossil records and ice cores from the poles demonstrates that climatic extremes far greater than previously envisioned by scientists and planners may well be experienced by humanity in the coming decades.

Climate change is essential to discussions of future water resource planning in the state—in spite of the fact that recent state water planning has minimized or ignored it. At the end of the day, knowing the uncertainties of global climate change will help us prepare more thoughtfully for the future. The good news in Texas is that there are water resources available to meet Texans' needs if we plan wisely and that we have a process in place that is transparent and welcomes public participation. It is up to us to stay informed and get engaged. The future of our children depends on it.

A dramatic rock formation towers over the convergence of the Pecos River and the Rio Grande. Photo by Wyman Meinzer.

2. THE MOLECULE THAT MOVES MOUNTAINS

The water molecule has several unique chemical properties that enable it to change the geologic structure of the landscape, control and alter the weather, and sustain life itself. As water moves on the Earth, it continually reshapes the landscape—cutting mountains, forming caves, and moving sediment to the ocean. At the same time the hydrologic cycle repeats its pattern of evaporation and precipitation in the form of rain, snow, and hail, leaving water to begin its journey again across the landscape to the sea. Along the way the water molecule sustains all plant and animal life.

WATER CHEMISTRY AND ITS RELATION TO GEOLOGY AND CLIMATE
Water is the only natural substance on Earth that is commonly found as a liquid, a solid, and a gas. Moreover, the unique properties of the water molecule enable it to affect both geology and climate. We can readily observe how water, by virtue of its physical and chemical characteristics, acts on the geologic makeup of

Karst, or limestone caves, show the dynamic effect water has on rocks as seen here in the Caverns of Sonora. Photo courtesy of the National Parks Service.

Texas by transporting sediments that reshape the landscape, including riverbeds and coastlines. Materials transported by water can range from large boulders to microscopic chemical elements. Water is called the universal solvent because of its ability to dissolve many substances. This property enables minerals to be carried from one area to another, for instance, from groundwater to surface water. This process can influence the formation of various geologic features, including remarkable stalactites and stalagmites deposited in caves and the fissures and pores in karst aquifers that allow extraordinarily rapid and unfiltered underground flow. Furthermore, the fact that water expands as it freezes creates massive forces that change physical geography by wearing away rocks, displacing soil, and even disintegrating mountains over time.

Climate is affected by the characteristics of water that allow rainfall, evaporation, clouds, fog, snow, hail, and ice. Water also influences weather by means of its thermal properties: its ability to retain temperatures moderates the climate. For example, along the Texas coast the air temperature might be in the 40s in the winter, while the ocean might be 60 degrees Fahrenheit (F) or higher. By the same process, in spring Gulf of Mexico water can be colder than the air temperature. Since oceans cover 71 percent of the Earth's surface, the thermal property of water is a stabilizing influence on the Earth's temperature. Temperatures on Earth can range from -112 degrees F to 136 degrees F. Compare this range to that of the moon, which has no water; its temperature ranges from -247 degrees F to 275 degrees F.

The water molecule is composed of two hydrogen atoms connected to one oxygen atom. The molecule is held together by the attraction of the opposite charges of the hydrogen and oxygen molecules. Because of these charges, the molecule can be described as a polar molecule, having a negative charge at the oxygen atom and positive charges at the hydrogen atoms. These charges attract other water molecules and play an important role in the properties of water. As water nears the freezing point, the molecules line up in a latticelike structure based on this polar attraction. This is why ice takes up more room than liquid water and can break pipes and rocks.

The reason ice floats on water is also explained by water's molecular structure. Ice is almost 10 percent lighter than liquid water because the molecules are farther apart in a structured grid in the frozen stage. The ability of ice to float is extremely important to the world's climate and plant and animal communities.

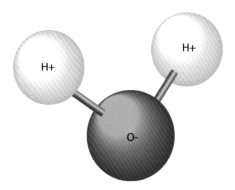

The water molecule. Image courtesy of the River Systems Institute.

Floating ice insulates the water below, which helps to maintain the temperatures of oceans, rivers, and lakes throughout the winter and enables plant and animal species to survive. About 75 percent of the freshwater in the world is in the form of ice.

Water also affects climate through the processes of evaporation and transpiration. Evaporation is the process by which water changes from a liquid state to a gas or vapor state. Transpiration is the evaporation of water from the leaves of plants. Evaporation from the oceans, lakes, and rivers provides almost 90 percent of the moisture in the atmosphere. Evaporation from the land and plants accounts for the other 10 percent. Together these two systems provide the moisture in the atmosphere that results in rain and snow and affect the overall climate.

In a state as large as Texas, we can feel the effects of varying atmospheric humidity on the average temperatures. For example, humid Houston in the southeast has an average summer temperature from 72 to 93 degrees F while dry El Paso varies from 63 to 95 degrees F in the summer—about a 50 percent increase in variability due to the thermal properties of water.

Capillary rise enables transpiration of water from its source in the soil through the plant leaves and into the atmosphere. Because water molecules stick to each other due to their polarity, they can literally climb up through the veins of a plant until they reach the leaves, carrying nutrients for the plant with them. This is the same capillary process that we see as paper towels soak up water. From the underside of the leaves, the water molecules are broken apart by the energy of the sun and evaporate into the air. Different plants evaporate water at different rates. An acre of corn gives off about 3,500 gallons of water a day, and a large oak tree can transpire 110 gallons of water a day. Invasive plants that are not native to an area often transpire greater amounts of water than native plants and can deplete the soil moisture and even affect streamflow. A single salt cedar or tamarisk, for instance, can transpire up to 200 gallons per day. Salt cedar dominates the riverbanks for miles along West Texas rivers and streams, including the Pecos and Rio Grande, and accounts for much of their reduced flow.

WATER AND GEOLOGY IN TEXAS

The surface configuration of Texas is closely linked to water and, in particular, by Texas' relationship to the sea. The existing Texas coast and sea level have been relatively stable for only 5,000 years—a mere tick of the geologic clock. Rocks near the town of Burnet in Central Texas tell us that seas covered much of the state approximately 400 million years ago. About 100 million years ago, a seaway reached from the Arctic Ocean to the Gulf of Mexico, covering all of Texas. At the end of the Cretaceous Period, about 65 million years ago, the Gulf of Mexico reached as far west as the Trans-Pecos region and north to the High Plains. Trans-Pecos refers to the area west of the Pecos River, bounded by the Rio Grande on the south and west, on the north by the thirty-second parallel, which forms the boundary with the state of New Mexico, and by the Edwards Plateau in the southeast.

Much geological evidence still exists from our undersea epochs. In the western part of the state, salt formations from remains of ancient seabeds affect the water quality of the Upper Brazos River and limit its uses. Visible formations of sandstone, shale, and limestone represent different depths of the ancient oceans, with limestone having been formed from the deeper areas. Sandstone formed along the shallow edges, and shale formed from medium depths. The colorful red rocks that can be seen in Caprock Canyon State Park, Palo Duro Canyon, and on El Capitan, one of Texas' highest peaks, were formed in an era of inland seas about 275 million years ago.

Texas and the Supercontinent

When looking at a physical geology map of Texas, there is a noticeable S-curve from Dallas–Fort Worth to Central Texas and on toward the Big Bend of the Rio Grande.

East of this line, which also closely follows much of the Balcones Escarpment, the formations generally run parallel. This line was the edge of the North American continent during the Cretaceous Period 60 million to 100 million years ago when dinosaurs roamed the Earth until their extinction 65 million years ago. The line closely follows the now-buried western portions of the Quachita Mountains, which are more than 300 million years old and are still seen in southern Oklahoma and near Marathon, Texas. The Quachita Mountains were formed about 250 million to 300 million years ago when the North American continent collided with Africa to form the supercontinent Pangaea, which is Greek for "all earth."

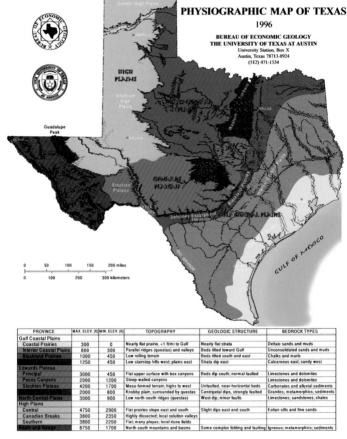

PHYSIOGRAPHIC MAP OF TEXAS
1996

BUREAU OF ECONOMIC GEOLOGY
THE UNIVERSITY OF TEXAS AT AUSTIN
University Station, Box X
Austin, Texas 78713-8924
(512) 471-1534

PROVINCE	MAX. ELEV. (ft)	MIN. ELEV. (ft)	TOPOGRAPHY	GEOLOGIC STRUCTURE	BEDROCK TYPES
Gulf Coastal Plains					
Coastal Prairies	300	0	Nearly flat prairie, <1 ft/mi to Gulf	Nearly flat strata	Deltaic sands and muds
Interior Coastal Plains	800	300	Parallel ridges (questas) and valleys	Beds tilted toward Gulf	Unconsolidated sands and muds
Blackland Prairies	1000	450	Low rolling terrain	Beds tilted south and east	Chalks and marls
	1250	450	Low stairstep hills west; plains east	Strata dip east	Calcareous east; sandy west
Edwards Plateau					
Principal	3000	450	Flat upper surface with box canyons	Beds dip south; normal faulted	Limestones and dolomites
Pecos Canyons	2000	1200	Steep-walled canyons		Limestones and dolomites
Stockton Plateau	4200	1700	Mesa-formed terrain; highs to west	Unfaulted, near-horizontal beds	Carbonates and alluvial sediments
	2000	800	Knobby plain; surrounded by questas	Centripetal dips, strongly faulted	Granites; metamorphics; sediments
North-Central Plains	3000	900	Low north-south ridges (questas)	West dip; minor faults	Limestones; sandstones; shales
High Plains					
Central	4750	2900	Flat prairies slope east and south	Slight dips east and south	Eolian silts and fine sands
Canadian Breaks	3800	2350	Highly dissected; local solution valleys		
Southern	3800	2200	Flat; many playas; local dune fields		
Basin and Range	8750	1700	North-south mountains and basins	Some complex folding and faulting	Igneous; metamorphics; sediments

Physiographic map of Texas, 1996. Image courtesy of University of Texas' Bureau of Economic Geology.

A pink granite batholith at Enchanted Rock State Natural Area. Photo courtesy of Texas Parks and Wildlife Department.

The Oldest Rocks in Texas

The oldest rocks in Texas form part of the core of the North American continent. These granite rocks were formed by volcanic activity 1.1 billion years ago and can be seen in the Trans-Pecos region and the area around Enchanted Rock State Natural Area in Central Texas near the town of Fredericksburg. Younger sedimentary rocks formed the uplifted mountains of the Trans-Pecos region as the seas receded.

The Formation of the Rocky Mountains and the Ogallala Aquifer

Between 40 million and 70 million years ago the Rocky Mountain region experienced a major episode of uplift. Toward the end of this violent era, West Texas was the scene of widespread volcanic and subterranean intrusion of granitic rocks. These rocks dominate in the Chisos and the Davis Mountains, but granite is also widespread in the Trans-Pecos region. The highest mountain in Texas is Guadalupe Peak (8,751 ft.) in Guadalupe Mountains National Park. The mountain range is the uplifted remains of an inland sea reef formation. Mount Livermore

(8,382 ft.) in the Davis Mountains is the second highest mountain in Texas and is volcanic in origin.

Beginning around 30 million years ago and continuing into the present, there has been renewed uplift, this time associated with the pulling apart of tectonic plates, producing the characteristic basin and range terrain of the Trans-Pecos region. "Basin and range" refers to the landforms characterized by long ridges and mountains with broad basins in between. Most notable of these is the basin and range of the southwestern United States, which occupies most of the area between the Colorado Plateau and the Sierra Nevada and from southern Idaho to Mexico.

At the time the basin and range region was forming, large river systems drained southeastward across Texas to the Gulf of Mexico, depositing sediment across the High Plains and simultaneously burying older marine deposits. Between one million and two million years ago, the Pecos River moved northward, cutting the Panhandle region off from its sediment source, resulting in the Caprock Escarpment dramatically severed by Palo Duro Canyon. These alluvial deposits formed the Ogallala Aquifer, which extends from Texas to Nebraska and is the largest aquifer in the United States and one of the largest in the world.

The Balcones Escarpment and the Edwards Aquifer

The Edwards Plateau, often called the Hill Country of Texas, rises generally north and west of the Balcones Escarpment, a 300-mile-long limestone ledge that trends southwest from Waco to San Antonio and then westward toward Del Rio. The exact cause and time of the uplift that created the plateau and the escarpment are not agreed on, but it is generally thought to have occurred between 10 million and 30 million years ago. The name Balcones is said to originate from the Spanish word *balcones* (balconies) and refers to the balcony-like appearance of the flat-topped hills that are seen across the hillside ramparts where the first native Texans stood to watch the arrival of the Spanish explorers.

Several rivers originating from springs of the Edwards Aquifer in the Hill Country cut through the plateau and the escarpment, forming dramatic canyons. These rivers include the Nueces, Frio, Medina, Guadalupe, and the Blanco. The Comal and San Marcos Rivers originate from springs that arise at the edge of the escarpment where the plateau meets the Blackland Prairies and the coastal plains.

THE GEOLOGY OF THE TEXAS COAST
Rivers and Coastal Plains

About 60 million years ago, the East Texas Basin was filling with sediments from the early Mississippi River, which flowed across East Texas and formed a delta north of what is now Houston. Although all of Texas was under the sea at one time, the rivers that met this Cretaceous sea and the inland sediment they carried helped to build the coastal plain. This sediment was laid down in a succession of ancient sloping deltas receding downward toward the coast. The arclike shapes of the coastal aquifers found as far up as the Balcones Escarpment essentially parallel the coast and follow these sediment wedge formations. Dumping of sediment has been building the coast for at least 65 million years and has extended the landmass of North America over 250 miles into the Gulf of Mexico. Today the Sabine, Trinity, Brazos, Colorado, and Rio Grande continue this process, although to a more limited degree as a result of the many dams built on them that trap much of the sediment upstream.

More recently the coastal plains have been influenced by the rise and fall of sea levels tied to changes in the climate. As the polar ice caps grow or shrink with changing temperature, they can hold more or less water in the form of ice. For example, during the peak of the ice age eighteen thousand years ago the Gulf of Mexico was as much as 400 feet lower than it is today.

The Texas Barrier Islands

Texas' magnificent barrier islands are only about five thousand years old—relative newcomers to the physiography of Texas. The largest of these, Padre Island, stretches 113 miles, making it the longest barrier island in the United States. During the last ice age when the sea level was lower than it is now, the coast was

Barrier islands line the length of the Texas Gulf Coast. Photo courtesy of Texas Parks and Wildlife Department.

about 50 miles farther out in the Gulf. As the sea level rose, the mouths of rivers were submerged, forming today's bays and estuaries. Sand that had previously been deposited at the rivers' mouths now lay underwater. Waves and storms reworked this sand and formed today's barrier islands. Sediment from rivers provided a ready supply of sand for these narrow islands, which today are eroding in places as a result of dam building upstream that has curtailed sediment flow to the Gulf.

Longshore drift is another force affecting the barrier islands; since the waves seldom strike the shore at a perfect 90-degree angle, there is usually a lateral movement of current so that sand is moved at an angle onto the beach. If all waves moved perpendicular to the shore, the beach would grow uniformly seaward. However, because the laterally moving current actually causes an angling of sand coming onshore, different parts grow at different rates. Because the Texas shoreline is curved, beach building occurs in two directions, and these converge centrally in Padre Island, one place on the Texas coast that is not eroding. This phenomenon is also the cause of the high deposition of litter on Texas' central coast.

Hurricanes and the Geology of the Texas Coast

Hurricanes continue to shape the Texas coastline. They cause more damage and geologic change from the resulting surge of water than from the direct effects of the wind. The increase in water height is a result of a combination of reduced atmospheric pressure and wind pushing the water in huge waves ahead of the storm. The architecture of the coastline is affected by hurricanes, especially in areas where they occur more frequently. The flat profile of Galveston Island and the Bolivar Peninsula, which are the most hurricane-prone areas of the Texas Gulf Coast, are to some extent examples of the reshaping of the coast by hurricanes. Notably, Galveston Island was hit in 1900 by the most devastating hurricane in U.S. history, which killed six thousand to twelve thousand people, the worst natural disaster ever to occur in the United States in terms of human life.

Landfalling hurricanes, 1950–2005. Data Source: National Oceanic and Atmospheric Association.

Since 1899 a significant hurricane has struck coastal Texas nearly every other year. In the 1930s and 1940s the hurricane average was higher, most likely due to a warming trend in those two decades, as hurricanes derive their strength from warm ocean waters. Since mid-1990 the state has experienced hurricanes in increasing numbers and intensity. The 2005 hurricane season proved a record year: twenty-eight named storms and seven major hurricanes. For the first time, two Category 5 hurricanes occurred in the Gulf of Mexico in one season: Katrina, which devastated New Orleans and the Mississippi coast, followed by Rita, which hit the Texas–Louisiana border. Hurricane Katrina was the most expensive natural disaster in U.S. history; Rita was the strongest hurricane to have ever entered the Gulf of Mexico.

To the surprise of many scientists, the 2006 season proved a welcome reprieve, with no major hurricane hitting U.S. shores. Scientists had predicted thirteen to sixteen named storms and eight to ten hurricanes for 2006. The reduced number of hurricanes was attributed in part to the early development of warm Pacific waters. This phenomenon, called El Niño, weakens Atlantic hurricanes and pushes them away from the East Coast of the United States.

HUMAN-CAUSED CHANGES IN TEXAS COAST GEOLOGY
Subsidence

Increasingly sophisticated human technologies, combined with the demands of the increasing population, have created unexpected consequences for the geology along the Texas coast. Overpumping of groundwater has caused the land surface to sink, in a process called subsidence, as much as 19 feet in areas of Galveston County. The impact of subsidence on increased flooding became apparent in 1961 when Hurricane Carla hit Galveston and low-lying areas were inundated. Other areas of the Texas coast have been affected by subsidence due to groundwater, oil, and sulfur pumping. As a result of the damage caused by subsidence, some of the first groundwater districts in Texas were formed to regulate the rate and amount of groundwater removal.

Dams, Jetties, and Beach Erosion

Construction for flood control, water supply, and navigation has caused accelerated beach erosion in communities such as Surfside, at the original mouth of the Brazos River. A negative effect of reservoir construction is that dams trap silt and sand that formerly would have replenished beaches washed away by Gulf storms. Portions of the coastline have been receding up to 10 feet per year, and the state is planning to spend millions in the Galveston and Surfside areas alone to rebuild portions of beaches lost to erosion.

The Brazos, the Rio Grande, and the Sabine are the three major Texas rivers carrying beach grade sand directly to the Gulf. Yet because there are numerous upstream dams on these rivers and reduced flows due to increased water use, the amount of sand delivered to the coast has been significantly reduced. The mouth of the Brazos was diverted in the 1920s to keep siltation out of the harbor at its mouth. This shortage of sand plays a major role in the current erosion rate of Texas beaches. Jetties con-

A jetty on South Padre Island. Photo courtesy of Texas Parks and Wildlife Department.

structed along the coast also obstruct the longshore drift, and harbor channels deepened for marine commerce alter the natural movement of sediment as well.

Global Warming and Rising Sea Level

More ominously today, increasing levels of carbon dioxide (CO_2) in the atmosphere and natural climatic cycles threaten to cause rising sea levels and profound changes in Texas weather patterns. In a reverse of the phenomenon that caused the sea level of Texas to be much lower during the Ice Age, melting polar ice caps—the result of global warming—are predicted to bring rising sea levels. Scientists project a rise in the sea level along the Texas coast of between 8 and 20 inches in the next one hundred years.

Most scientists agree that global warming is not solely the result of a natural cycle but is accelerated by major increases in the burning of fossil fuels, which injects ever greater amounts of CO_2 into the atmosphere—more than a 30 percent increase since the years prior to the industrial revolution. Scientists have not seen CO_2 levels this high for at least 420,000 years. By analyzing tree rings, ice cores, and sediments, scientists have been able to determine CO_2 levels as far back as several hundred thousand years. Ancient changes in CO_2 levels are attributed to various major phenomena, such as large volcanic eruptions, changes in the flow of the Gulf Stream, and even changes in the Earth's orbit—all occurrences beyond the control of humans. Today the general consensus among scientists is that current elevated levels of CO_2 are largely attributed to the actions of humans and are a principal contributor to global climate change.

Rising sea levels have many consequences, including increased flooding, an increase in the salinity of groundwater and surface water near the coast, and loss of wetlands. In the future, residents with houses on the coast will likely be forced to move them or tear them down. Sadly, continued erosion and potential sea level rise simultaneously reduce the expanse of wetlands in the bays and estuaries of the Texas coast. Healthy, abundant wetlands are invaluable and play critical roles that we often fail to recognize:

- **WATER FILTERING:** Plants in wetlands absorb many of the wastes that humans discharge into rivers.
- **NURSERIES:** Nearshore wetlands are the nurseries for many species of fish, crabs, and shellfish.
- **BIRD HABITAT:** The Texas coast is on the major North American flyway for migratory birds, many of which spend the winter on the coast, helping our state to have greater bird diversity than any other, with the possible exception of Hawaii.
- **FLOOD BUFFERS:** Wetlands reduce the severity of floods by serving as detention facilities. Without wetlands, flooding increases. Wetlands help to absorb the energy of hurricanes, reducing the storm surge. Most experts believe the power of Hurricane Katrina would have been reduced had there not been major alterations and reductions to wetlands between New Orleans and the open Gulf on the Mississippi River delta.
- **EROSION CONTROL:** Nearshore wetlands act as buffers, reducing shoreline erosion.
- **RECREATION:** Wildlife observers in Texas spent $1.3 billion in 2001, much of that on birdwatching, which is dependent on wetland habitat. Hunters and anglers spent another $3.5 billion, with much of this activity focused on the coast.

TEXAS CLIMATE
The Gulf Dominates
Most of Texas has a modified marine climate; that is, it is dominated by the onshore flow of tropical ocean air from the Gulf of Mexico. Between October and June, the tropical moisture collides with colder air from the north, resulting in thunderstorms over Central Texas as the Balcones Escarpment causes incoming moist Gulf air to rise and cool. Texas is a battleground between warm moist air from the tropics and cool dry air from the Great Plains. Along the Gulf Coast, hurricanes can bring large rain events in short periods of time, including flooding from storm surge and damage from high winds. The peak time for hurricanes is August and September, but the season extends from June to November.

Contrasting Geography and Climate

Because of the sheer size and topographic diversity of Texas, the climate varies greatly from north to south and east to west. The Great Plains, which experience sweeping cold fronts from Canada, lie to the north. Splitting the state is the one-hundredth parallel, the north-south meridian lying between areas to the east that could reliably support farming without irrigation and lands to the west where rain is sparse and sporadic and where ranching is the most common land use. To the east is Louisiana, one of the nation's wettest regions. Farther west is the searing heat of the Chihuahua Desert. In addition to the Gulf of Mexico, the Pacific Ocean provides moisture to the Texas climate as it is drawn north over Mexico. In 1998 Hurricane Madeline in the Pacific off of southern Mexico brought additional moisture that caused record flooding in parts of Central Texas.

Texas Climate Patterns and Extremes

RAINFALL AND FLOODS Although the rapid changes in our weather are legendary, there are general patterns of climate in Texas amid the extremes. These patterns often show up as averages—rarely resembling measurements actually seen at any one time.

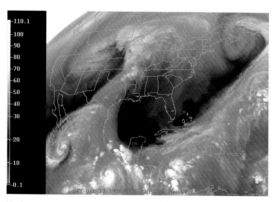

Pacific Hurricane Madeline caused record flooding in Central Texas in 1998. Photo courtesy of the University Corporation for Atmospheric Research (UCAR).

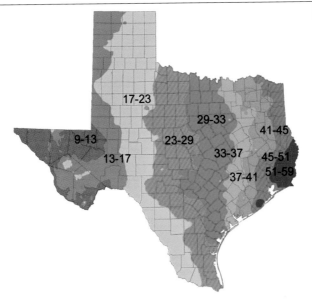

Precipitation in Texas. Data Source: Texas Parks and Wildlife Department.

Average annual rainfall trends from east to west, with Orange in the Sabine Valley averaging about 59 inches per year and El Paso at the far western tip averaging about 8 inches. This east-west precipitation trend is part of a larger pattern that comes off the Great Plains. The wet season is not the same statewide. North, Central, and East Texas receive their maximum precipitation in May, whereas the High Plains and Trans-Pecos receive their peak rains in the warmest time of year, between May and October. Along the Gulf Coast, late summer and early fall comprise the wet season.

One climatic or hydrologic extreme that distinguishes Texas is flooding. Texas is number one on the world rainfall charts for the twenty-four-hour record of 43 inches of rainfall at Alvin in 1979, associated with Tropical Storm Claudette. Texas also holds the U.S. twelve-hour rainfall record of 32 inches at Thrall in Central Texas in 1921. In 1935, 22 inches of rain fell in two hours and forty-five minutes at D'Hanis near Uvalde, west of San Antonio.

The Pedernales River is a torrent during a flood. Photo courtesy of Texas Parks and Wildlife Department.

The worst flood in terms of fatalities occurred in 1913 on the Brazos River. This flood killed 177 persons, and the high number of fatalities was attributed to the building of levees for flood protection. As a result of a larger previous flood on the Brazos in 1899 that killed 35 persons, numerous levees were built along the river for protection. The levees caused the water to rise even higher before it overflowed into the former floodplains. Tragically, this created an unexpected sudden rush of water in which many were trapped and subsequently drowned.

Prior to Hurricane Katrina, Tropical Storm Allison, in 2001, was the costliest urban flood in U.S. history, inundating seventy thousand buildings and causing $5 billion in damage in the Houston area from its 30 inches of rainfall. Perhaps most alarming, Allison's flooding shut down the Texas Medical Center, a major health complex and emergency-care facility of two medical schools, thirteen hospitals, and over six thousand beds. Ironically, a good portion of the flooding from Allison was due to flood control efforts to straighten, widen, and line area streams with concrete, which resulted in faster and greater runoff and flooding. In addition, the area from Austin to San Antonio is one of the nation's three most flash-flood-prone regions.

TEMPERATURE Unlike the east-west rain patterns, temperature patterns trend south to north, from warmest to coldest. The winter mean daily minimum temperatures are in the upper teens in the Panhandle, while in the Lower Rio Grande Valley the mean minimum temperature is in the low fifties, enabling winter citrus production in most years. Winter highs vary from the upper forties in the north to almost seventy in the south. Summer low temperatures average in the low sixties in the Panhandle to the mid-seventies in the south, while highs reach the nineties in both regions. Today, as most Texans know, high temperatures can reach the hundreds in most parts of the state.

In fact, the highest temperature recorded in the state was 120 degrees F at Seymour, near Wichita Falls, in 1936. In 1980 Wichita Falls recorded ten consecutive days over 110 degrees F. Occasionally it can get cold in Texas. In 1899 at Tulia, a small town between Lubbock and Amarillo, the temperature dropped to -23 degrees F.

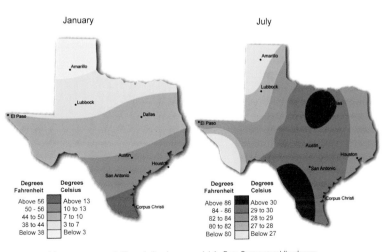

Mean temperatures in Texas during January and July. Data Source: worldbook.com.

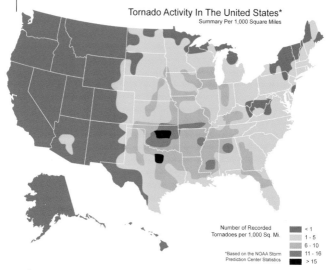

Tornado activity in the United States. Data Source: National Weather Service—Storm Prediction Center.

Water plays a role in temperature. In drier areas of the state, the daily swings and seasonal cycles in temperature from day to night and summer to winter are greater in contrast to the more humid areas, where the air's water content mutes these shifts.

TORNADOES Unfortunately, another of Texas' lethal climatic extremes is the tornado, which usually forms in conjunction with violent thunderstorms. Texas experiences more frequent tornado activity than any other state. It is the largest state in Tornado Alley, which contributes to this dubious distinction. The United States has ten times more tornadoes than any other country in the world, and there is an increasing trend in the number of tornadoes both in Texas and in the United States. The list of major tornadoes in Texas is long, but a few stand out—the May 1953 tornado that virtually walked 5 miles through Waco and killed 114 persons and the 1997 F-5 tornado in Jarrel, north of Austin, that resulted in 27 deaths. An F-5 is the severest rating for tornadoes with winds from 261 to 318 miles per hour, exceeding even the most powerful hurricane winds.

HAIL Although Texans have been known to pray for rain, we sometimes get falling moisture in an unwelcome, unfriendly form—hail. If Texas lies in Tornado Alley, it is also located in the Hail Expressway. Tornadoes and hail form from similar storm patterns as rising warm air meets descending cold air. Raindrops are frozen and refrozen as they fall and are tossed back up into the colder air repeatedly until they get large enough to fall through the storm. The size of the hail depends on the number of times the hailstone goes through this process. Softball-sized hail is not unheard of in Texas. In 1917 in Ballinger, near Abilene, hail covered the ground up to 3 feet deep.

The velocity of falling hail can cause more damage, even with smaller hailstones. The maximum falling vertical velocity of a

Hail Days Per Year (1980-1994)

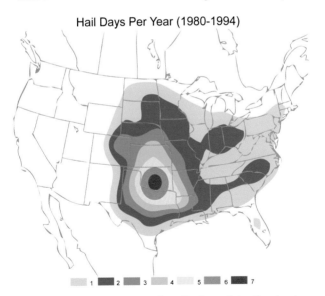

Number of hail days per year in the United States. Data Source: National Oceanic and Atmospheric Association.

3-inch hailstone is 90 miles per hour, but with a downdraft or even side winds, the velocity can increase. Of course, chunks of ice traveling 100 miles per hour not only damage property but also injure or kill humans and wildlife. A large hailstorm hit Fort Worth in 1995 with baseball-sized hail, killing one person and injuring a number of others.

DROUGHT Droughts do not come as quickly and strike as hard as floods and tornadoes. Instead, they sneak up on us and cause more economic damage than major storms. A brief drought in Texas in 1996 caused agricultural losses estimated at $2 billion. Unfortunately, drought in Texas is a normal condition. A Texas Water Commission (now Texas Commission on Environmental Quality) study found that it is more likely that a six-month to one-year drought will occur somewhere in Texas than average precipitation during the same period. In the years ahead, this trend is likely to increase due to global warming.

One reason that Texas is so drought-prone is its latitude, the same latitude as the Sahara Desert. As in the Sahara, large high-pressure cells can sit over the state for weeks or months at a time and block storms and incoming moisture. The cause of these cells is not well understood but possible influences are solar storm cycles, ocean temperature cycles (El Niño and La Niña), and global warming.

Furthermore, scientists are finding evidence that the droughts of the twentieth century are not nearly as severe as the droughts hundreds of years ago. A twenty- to fifty-year drought in the southwestern United States in the 1600s may have caused the abandonment of Indian pueblos. Prior to the 1600s, droughts lasted for one or more decades, not just a matter of years, like the pattern we have seen for the past four hundred years. Most Texans consider the drought of record to have occurred from 1950 to 1956. Of the 254 counties in the state, 244 were declared disaster areas. Comal Springs, in New Braunfels, the largest spring in Texas, stopped flowing for almost five months. However, in the 1990s parts of far West Texas and South Texas had droughts that rivaled that of the 1950s.

HUMANS—PROBLEMS AND SOLUTIONS

The only consistency is that the climate of Texas is inconsistent. Extreme forces such as the rise and fall of sea level, hurricanes, droughts, and floods have shaped, and will continue to shape, the physical geography of the state and its future water supplies. Extreme weather in the form of rain events associated with hurricanes and violent thunderstorms provide much of our water resources—or in the case of drought, challenge our water resource planning. In addition to these natural forces, human actions are causing alterations in the geology and climate of Texas unprecedented in its natural history.

Because water plays a significant role in the geology and climate of Texas, how we manage water can affect the physical geology of the state as well as our ability to endure climatic extremes. The damage from hurricanes can be reduced by preserving wetlands through beach erosion control and by allowing sediment transport from adequate river flows. By wisely using groundwater, subsidence can be prevented and spring flows can be maintained, allowing humans and aquatic ecosystems to withstand droughts. Thus humans have the power and ingenuity to correct some of our negative impacts on natural systems and, ultimately, to mitigate some of the damage we have caused.

Morning light creates a glow on the Upper Sabinal River. Photo by Wyman Meinzer.

3. A TEXAS WATER JOURNEY

Some 2,500 miles of Texas' borders are formed by the rivers of Texas and the Gulf of Mexico. If you add the wanderings of the tidal shoreline, that number would be 5,500 miles. The Rio Grande in the west, the Red River in the north, the Sabine River in the east, and the Gulf Coast all define the shape and heritage of Texas. Like the meandering of river channels, the boundaries have changed as various groups have explored, claimed, fought over, and purchased the state we know today as Texas.

Many of us associate Texas with drought, and this is reasonably true; much of the state is never very far from being parched. But Texas has 15 major rivers and 3,700 named streams that wind over 191,000 miles. Many of these rivers and streams provide needed freshwater for the seven major estuaries. With 212 major reservoirs and more than 5,000 smaller ones, Texas, surprisingly, is second only to Minnesota in surface area of lakes and reservoirs and second to Alaska in volume of inland water.

These lakes and reservoirs can hold about 60 million acre-feet. The total average runoff, or streamflow, from the state is 49 million acre-feet per year.

Springs are another aquatic surprise for those who think of Texas' landscape as just cactus and mesquite. Texas has an abundance of springs—hundreds of them, some forming rivers. The largest artesian spring west of the Mississippi River is Comal Springs in Central Texas, and just 16 miles away is the second largest artesian spring system west of the Mississippi at San Marcos. Hundreds of miles west, along Interstate Highway 10 at Balmorhea State Park, is the largest spring-fed swimming pool in the United States. Springs emerge from aquifers, and 76 percent of the state is over an aquifer. Combined, Texas' 9 major and 21 minor aquifers hold an estimated three to four billion

Major Texas rivers and basins. Data Source: Texas Parks and Wildlife Department and Texas Water Development Board.

gallons of water; however, only about 10 percent of this ground-water is recoverable with existing technology.

We are about to embark on a journey down the rivers of Texas from their origins to the coast, where they provide crucial freshwater inflows for bays and estuaries. Rather than explore individual rivers that slice through all parts of the state, we will examine them as parts of large geographic areas. By using this method, we can look at river segments that share similar geology, geography, climate, demography, and land use, as well as history. We will not only examine rivers and reservoirs but also the aquifers and springs that in many instances provide water for agriculture and municipalities and sustain the rivers, especially during times of drought. We begin the journey in the Rocky Mountains of southern Colorado at the headwaters of the Rio Grande.

The Upper Rio Grande and Trans-Pecos region. Data Source: Texas Parks and Wildlife Department and Texas Water Development Board.

THE UPPER RIO GRANDE AND THE RIVERS
OF THE TRANS-PECOS REGION
The Rio Grande—Not Just a Texas River

The river most closely associated with Texas, the Rio Grande, is not the exclusive property of Texas but is shared with two other U.S. states and five states in Mexico. In Mexico it is known as the Rio Bravo. The fact that the Rio Grande is so readily connected to Texas is probably related more to history than geography or geology.

The origin of the Rio Grande basin and its contributing watersheds is in southern Colorado. The Rio Grande is the second longest river in the United States, and most of its length borders Texas. During the time Texas was an independent nation,

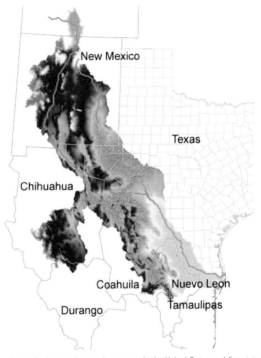

The Rio Grande Basin extends over three states in the United States and five states in Mexico. Data Source: River Systems Institute.

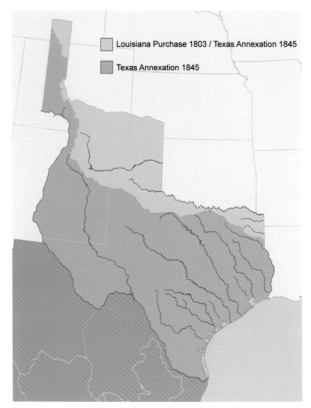

Texas boundaries in 1845. Data Source: National Atlas of the United States.

1836–1846, the western boundary included the Rio Grande all the way to its headwaters and beyond to what is now Wyoming. The boundary close to what we see today was established by the Treaty of Hidalgo in 1848, which ended the Mexican War and firmly recognized the annexation of Texas to the United States, which had occurred two years earlier. As early as 1721 the Medina River in Central Texas and as late as 1811 the Nueces River west of the Medina were considered the western boundary of what was then a territory of Mexico.

The Rio Grande was never explored in its entirety in a single expedition. Coronado is said to have crossed the Rio Grande in 1540. Several other Spanish adventurers crossed it in the vicinity of El Paso as they followed the Camino Real, which ran from Mexico City to Santa Fe. Another old Spanish trail, El Camino Real de los Tejas, connected Mexico City with Nachitoches, Louisiana, and crossed all major Texas rivers.

It seems fitting that the oldest town in Texas, Ysleta, which is now part of El Paso, became part of the United States only because of the shifting channel of the Rio Grande. Established in 1680 on the Mexican side, Ysleta found itself marooned on an island in the Rio Grande after floods in 1829–1831. When the United States annexed Texas, the deepest channel was declared the border, putting Ysleta in Texas. Disputes over the western boundaries of Texas have continued and necessitated the establishment of a permanent agency, the International Boundary and Water Commission, to adjudicate boundary issues. Channel changes in the Rio Grande have caused disputes for years, especially between the cities of Juarez and El Paso where land has been ceded from one country to the other. Today the river channel between the two cities is lined with cement to avoid the shifting of boundaries; the Chamizal, a site maintained by the National Park Service, lies in the no-man's-land between.

THE RIO GRANDE BELOW EL PASO—SEARCHING FOR THE LAST DROP The amount of flowing water that Texas receives in the Rio Grande at El Paso is determined by nature to some extent. However, legal agreements have largely affected the river's flow. Snowfall in the Rockies and Sangre de Cristo Mountains of Colorado and New Mexico and infrequent rain in the deserts of New Mexico are diverted or stored behind numerous dams in New Mexico, including the Elephant Butte Dam near the Texas border. Interstate agreements mandate irrigation releases for Texas and Mexico, as well as Colorado's obligatory releases upstream, but they are not sufficient to sustain flow in the river below El Paso year-round.

There are no major tributaries of the Rio Grande for 350 miles between Elephant Butte Reservoir and Presidio, Texas, which is 200 miles downstream of El Paso. This stretch of the river is known as the Forgotten Reach because it has low or no flow, and few humans inhabit its shores. In fact, because of the limited surface water available from the Rio Grande, El Paso relies mainly on groundwater sources. The reduced flows also mean less dilution of contaminants. Heavy metals and pesticides from industry and agriculture have been identified in the Rio Grande. Untreated wastewater from Mexico has contributed to elevated fecal coliform levels downstream of major cities like Ciudad Juarez and smaller cities including Ojinaga, which is located across from Presidio.

THE RIO CONCHOS—REBIRTH OF THE RIO GRANDE In truth, today there are two Rio Grandes due to the virtual absence of natural flow for about 200 miles below El Paso. Upstream from the city of Presidio and Ojinaga, the Rio Conchos, which is the largest Mexican tributary, joins the Rio Grande and contributes almost 75 percent of the flow at that point.

The Rio Conchos flows help to maintain recreational use of the Rio Grande through miles of Big Bend Ranch State Park, the spectacular canyons of Big Bend National Park, along the Black Gap State Wildlife Management Area, and another 127 miles of designated Wild and Scenic River below the park known as the lower canyons. Hot springs along the shore flow into the Rio Grande near Boquillas Canyon, where a thriving bathhouse stood until the 1920s. Unfortunately, continued droughts and increasing water diversions both in Mexico and in the United States have reduced the flows of the river to the point that river recreation is not always possible. At various times since 2000, the river has stopped flowing altogether in Big Bend National Park. These are the first cessations of flow in the park since the 1950s drought.

Major reservoirs and diversions for irrigation on the Rio Conchos in Mexico have resulted in periods of low flows that at times do not meet the flow agreements between the United States and Mexico. According to a 1944 treaty, one-third of the annual flow from the six Mexican tributaries is supposed to reach the Rio Grande. Having fallen behind in water payments

Big Bend Ranch State Park is a spectacular combination of desert and river ecology. Photo courtesy of Texas Parks and Wildlife Department.

resulting from droughts at the start of the new millennium, Mexico repaid the water debt through releases in early 2005 after the drought eased.

THE PECOS AND DEVILS RIVERS AND LAKE AMISTAD Above Del Rio, the Rio Grande receives some much-needed flow from both the Pecos

River and, after several miles, the Devils River. The Pecos begins at 13,000 feet in the Sangre de Cristo Mountains east of Santa Fe, New Mexico, and flows for over 900 miles to its confluence with the Rio Grande, draining 44,000 square miles. Yet, like the Rio Grande, the Pecos is subject to much diversion and reservoir impoundment in New Mexico before flowing hundreds of miles through the driest parts of Texas. Salt cedar or tamarisk, an invasive plant introduced in the 1800s for erosion control and as a windbreak, has choked miles of Pecos and Rio Grande riverbanks, increasing the already high evaporation rates. Tamarisk has become such a problem that in late 2006 the federal government passed a tamarisk control bill authorizing millions of dollars in grants and public-private partnerships for control and eradication of tamarisk throughout the western United States.

In comparison to the long journey of the Pecos, the Devils River travels only approximately one hundred miles after arising from springs in the Edwards Plateau. Although the area along the Lower Pecos and the Devils River is virtually uninhabited today, prehistoric peoples moved into the area twelve thousand years ago, living in caves along the canyon walls above the Rio Grande, Pecos, and Devils Rivers. About seven thousand years ago, they began a unique tradition of spiritual rock art, much of which remains on cave walls today. Seminole Canyon State Park preserves several of these sites.

Taking advantage of the convergence of these three rivers and the deep canyons, engineers chose the site for what is now Lake Amistad. The lake, whose name comes from the Spanish "friendship," was a cooperative endeavor between the United States and Mexico and was completed in 1969. It provides storage of water for both countries. Amistad is the second largest lake in Texas, but it suffered greatly from the droughts of the late 1990s, its level dropping to less than 20 percent of capacity.

Construction of the lake inundated several archeological sites, some of which were excavated before it was filled. Many caves with lovely prehistoric rock art are now underwater. Other caves that were high above the natural river are accessible only by boat; these caves are under protective jurisdiction and

An archaeological excavation prior to the flooding of the Amistad Reservoir. Photo by E. Mott Davis, courtesy of www.texasbeyondhistory.net.

must be visited with guides. The remains of a railroad tunnel built in 1877 for the San Antonio–El Paso route that climbed from the riverside to the top of the canyon walls lies partially submerged at the edge of the lake. Formerly known as the third largest spring in Texas, Goodenough Springs is also submerged beneath Lake Amistad. These springs are still strong enough to flow out at the bottom of the lake. Although the springs are about 150 feet below the surface, their warmer temperature is noticed in the winter at the surface of the lake.

THE RIVERS, AQUIFERS, AND PLAYAS OF THE HIGH PLAINS REGION
The Red and Canadian Rivers

Like the Rio Grande and the Pecos, the Canadian and Red Rivers originate in the Sangre de Cristo Mountains. However, these rivers trend eastward and cross the High Plains. The Red River originates in eastern New Mexico and eventually joins the Mississippi 45 miles northwest of Baton Rouge, Louisiana. It is the second longest river associated with Texas and since 1836 has formed part of the state's northern boundary with Oklahoma. The Canadian River also starts in eastern New Mexico and then cuts across the Texas Panhandle for 170 miles on its way to meet the Arkansas River in Oklahoma.

A tributary of the Red River carved the famous Palo Duro Canyon explored by Coronado in 1541. In the 1700s French traders from East Texas journeyed to the Upper Red River to trade with the Indians. In 1806, in a wave of scientific discovery following the success of the Lewis and Clark expedition, the United States sponsored an expedition led by the astronomer Thomas Freeman and the naturalist Peter Custis to explore the Red River and locate a possible water route to Santa Fe. This expedition was considered second in importance only to the Lewis and Clark expedition because it would also test the Louisiana Purchase's disputed border with Spain. Spain, on hearing of the plan, sent an armed contingent that caused the expedition to turn back after 615 miles.

The Texas High Plains region. Data Source: Texas Parks and Wildlife Department and Texas Water Development Board.

The Lighthouse at Palo Duro Canyon State Park. Photo courtesy of Texas Parks and Wildlife Department.

High Plains Water Use

Water use in the High Plains is mainly for agriculture and comes principally from the Ogallala Aquifer. Region A, the state water planning unit responsible for the northern part of the Texas Panhandle and the High Plains, is one of the largest water-consuming regions in the state even though it is sparsely populated. It accounts for about 13 percent of the state's water use and only 1.9 percent of the population. Although Amarillo, with a population

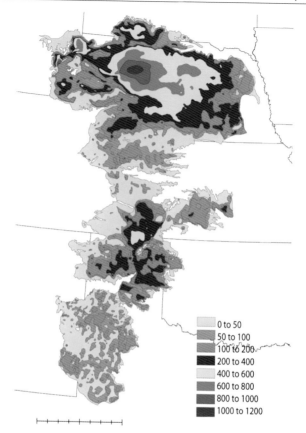

	0 to 50
	50 to 100
	100 to 200
	200 to 400
	400 to 600
	600 to 800
	800 to 1000
	1000 to 1200

Saturated thickness of the Ogallala Aquifer in feet in 1997. Data Source: United States Geological Survey and Texas Commission on Environmental Quality.

of 173,000, is the only major city, 49 percent of the corn and 74 percent of the swine raised in Texas come from Region A.

The Ogallala Aquifer is one of the largest aquifers in the world, covering 174,000 square miles from South Dakota to Texas. The Texas portion of the Ogallala contains approximately 417 million acre-feet of water.

Water use from the Ogallala exceeds the rate of replenishment. Unfortunately, the Ogallala recharges very slowly, and what little recharge there is comes mainly through small depressions in the plains called playa lakes. Between 1940 and 1980 the aquifer's average water level dropped nearly 10 feet per year; in the 1990s the rate slowed to less than 1.5 feet per year. In the 2007 State Water Plan, the 2060 projected decline from the 2010 level is 52 percent. The good news is that the depletion of the Ogallala is slowing, but it still outpaces the recharge. Two factors are slowing this depletion: reduced acreage under irrigation and more efficient water use. Irrigated acres on the Texas High Plains dropped from 3.95 million in 1979 to 2.7 million in 1994, in part as a result of lower aquifer levels and increased pumping costs, primarily for energy. Irrigation water efficiency is measured as the percentage of water delivered to the soil compared to the amount originally pumped. Water efficiency for irrigation increased to 75 percent in 1990 from about 50 percent in the 1970s. The more recently used low-pressure, full-dropline center-pivot irrigation systems have close to 95 percent efficiency; buried drip lines are close to 100 percent efficient.

Playa Lakes

Playa lakes provide a significant portion of the minimal recharge to the Ogallala Aquifer. They are ephemeral ponds that are filled in a variety of ways, including rainwater and runoff from irrigation or from cattle watering. The western Great Plains of the United States contain over forty thousand playas, accounting for more than 95 percent of the playa lakes in the world. Although the Texas High Plains receive less than 20 inches of rain per year, this area contains an estimated twenty thousand playa basins. The watershed for these depressions in the landscape can vary from a few to thousands of acres. Some playas are salt lakes and receive water from Ogallala seeps or springs.

Playa means "beach" in Spanish and may refer to the formation's sandy shorelines. There is disagreement about how playas are formed, but wind and land subsidence are the two most common theories. Ancient Indians camped beside some of the major playas, not only as a source of water, but also to take advantage of the abundant wildlife, including waterfowl, that de-

Aerial view of a few of the 20,000 playa lakes in the Texas High Plains. Photo courtesy of Texas Parks and Wildlife Department.

pend on these wetlands. Today the playas are critical habitat for over one million ducks and geese. Early Spanish explorers also wrote about using playas for their campsites. In the early 1900s bankers used the water levels of playas to assess rainfall and adjust loans to farmers accordingly.

Farming and irrigation are affecting the condition of playas. Farming practices often result in siltation of playas, reducing their capacity to recharge the Ogallala. At the same time, more efficient irrigation practices and the reduced acreage of irrigated land have lowered the amount of excess runoff that once provided extra water for playas. Playas themselves are used as sources of irrigation water; they provide 10 to 25 percent of the annual irrigation supplies in some counties of the western Great Plains. The Playa Lakes Joint Ventures (PLJV), a major conservation partnership, estimates that playas occupy 2 to 5 percent of the land surface of the western Great Plains but are responsible for 85 to 90 percent of the recharge of the Ogallala. Recharge under a playa can exceed 3 inches per year, while between playas the rate occurs at 0.004 inch per year.

Because of the playas' importance for aquifer recharge, as wildlife habitat, and as sources of irrigation water, the PLJV was formed in 1989 to assist in playa conservation. It has raised more than $50 million to conserve playas and promote responsible land practices around playa lakes.

The Upper Colorado and Upper Brazos Rivers

Two of Texas' major rivers, the Colorado and Brazos, originate in the High Plains. Their flows are often quite low in the upper reaches because of the low rainfall and the absence of significant springs. The Colorado is the longest river with a drainage basin entirely in Texas. The Brazos River, the third largest in the state, originates in Texas, but its drainage basin includes parts of New Mexico. These two rivers flow generally southeastward to the Gulf of Mexico.

Both the Colorado and the Brazos face significant problems from natural and man-made salt pollution in the upper reaches. Natural salt contamination can come from exposed salt formations in the riverbeds; man-made salt contamination can come from oil drilling. Total dissolved solids (TDS) is a measurement of salinity. The salinity of seawater is 35,000 milligrams per liter (mg/L), whereas freshwater is defined as having less than 1000 mg/L TDS. The average TDS for Salt Croton Creek on the Upper Brazos has been measured at 71,237 mg/L, or more than twice that of seawater. Drinking water from the three main-stem reservoirs on the Brazos—Possum Kingdom, Lake Granbury, and Lake Whitney near Waco—has to be treated with expensive demineralization processes, including reverse osmosis. High salinities in these Brazos lakes have also been linked to outbreaks of golden algae, which, although not known to be harmful to humans, has resulted in fish kills of up to four million fish in spring 2005. On the Upper Colorado, diversion systems and evaporation reservoirs have been constructed to dilute the effects of salty normal river flows. At low flows, the higher chloride water is removed from the river channel and pumped to the off-channel reservoirs where it is evaporated.

THE RIVERS OF THE NORTHEAST AND EAST TEXAS REGION
The Red River and Lake Texoma

The Red River flows out of the High Plains of the Panhandle and becomes the northern boundary of Texas. However, due to the continual shifting of the river's channel, the Red River Boundary Commission was created in 1995 to monitor the boundary between Oklahoma and Texas. Lake Texoma sits along this boundary and is one of the largest reservoirs in Texas, with a storage capacity of 4,505,000 acre-feet. Lake Texoma is shared with Oklahoma and suffers from natural salinity problems from upstream salt deposits and saline springs. Measures are being taken to reduce the chlorides causing the salinity, including a 340-foot-diameter dike around Estelline Springs to control its flow into the Red River. In the 1950s the city of Dallas, in desperate need of additional water, built an emergency pipeline to Lake Texoma only to find that the water was too salty to use. Now, fifty years

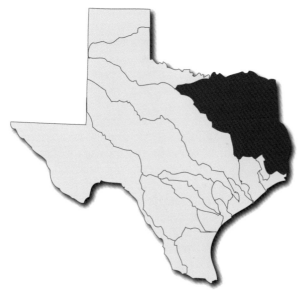

The Northeast and East Texas region. Data Source: Texas Parks and Wildlife Department and Texas Water Development Board.

The Wichita River is a major tributary of the Red River. Photo by Angeline M. Van Zant.

later, Dallas is again considering using Lake Texoma water. By using reverse osmosis to eliminate salt, water can be used by municipalities, but it is a very costly alternative water source.

Since the 1830s steamboats plied many Texas rivers, including the Red; however, a 75-mile-long logjam obstructed the river upstream of Shreveport. After the logjam was cleared in 1834, the Red served as the most important Texas river for navigation. The logjam returned again in 1856 and raised the water level in Caddo Lake, backing up water to the town of Jefferson. Jefferson became the major river port of Texas until 1874, when the logjam was removed.

Caddo Lake

Caddo Lake was one of the largest natural lakes in the southern United States until it was dammed in 1914 just over the Louisiana border. This shallow lake still has the feel of a natural lake, with a labyrinth of thousands of cypress trees, some of which are more than five hundred years old. The lake is named after the Caddo Indians, which included twenty-five distinct groups sharing the same language and inhabiting an area along the Red

A logjam on the Red River in 1873. Photo courtesy of the Noel Memorial Library, Louisiana State University in Shreveport.

River where Texas, Oklahoma, and Arkansas come together. The Caddo were relatively peaceful, agricultural people who lived in small villages. In the late 1680s Spanish soldiers and priests settled among the Caddo and tried to organize them into towns, but the Caddo resisted conversion to Christianity and larger settlements. The last existing Caddo Indian village was on the shores of the lake.

The logjam on the Red River in the mid-1800s raised the level of water and enabled steamboats to come from the Mississippi River to Jefferson, which was only a few miles upstream from Caddo Lake. Jefferson ranked second only to Galveston in volume of commerce during its thirty-year reign as Texas' chief river port. When the lake level dropped after the removal of the logjam, a new industry sprang up: freshwater pearls. This boom ended in 1914 when the dam was built in Louisiana, raising the lake again and making the pearls inaccessible to hand digging. Other luxuries were drawn from the waters of Caddo Lake: roe from the now-endangered paddlefish were shipped in refrigerated railroad cars to the East Coast and sold as caviar. The next commercial boom was oil in the early 1900s, and Caddo Lake

Paddlers weave in between trees on Caddo Lake. Photo courtesy of Texas Parks and Wildlife Department.

had the first offshore drilling rig in the United States. Oil continues to be pumped from the lake today.

Today Caddo Lake is world famous as a unique habitat for resident and migratory wildlife, home to a state park, a wildlife management area, and a national wildlife refuge. It has been designated a wetland of international importance by the United Nations—one of only a few sites in the United States. Ongoing disputes over sales of water from the lake began in the early

2000s, and continued water appropriations threaten its fragile ecosystem. Controversies have also arisen over mercury contamination of the lake, evidently linked to airborne deposition from power plant emissions.

Dallas and Fort Worth Share the Trinity River

The Trinity River has several forks that start less than 200 miles from the Dallas–Fort Worth area, converging in and around the metropolitan area. The Dallas–Fort Worth Metroplex is the largest inland U.S. metropolitan area. As a result, the Trinity River in this area has been transformed from its natural state by diversions, wastewater, and runoff from urban streets. The river in this area receives approximately 550 million gallons per day of domestic municipal effluent.

Planning Region C, which includes Dallas–Fort Worth, had a total population of 5,254,722 in 2000, or 25.2 percent of Texas' population. Most of Region C is in the upper portion of the Trinity Basin. Interestingly, this region accounts for only 7.2 percent of the state's water use—a result of only minor water usage for agricultural irrigation. Eighty-five percent of the region's water use is for municipal supply. In fact, the per capita use is highest among the state's metropolitan areas—as high as 260 gallons per person per day, compared to 220 in Austin, 142 in San Antonio, and 137 in El Paso. Manufacturing and steam electric power are the next largest water users in the Metroplex.

Surface water, mainly from the thirty-four reservoirs in this region, supplies over 90 percent of the region's needs. Surface water imported from other regions is another significant source of water for the Metroplex. In its 2007 proposed water plan, Region C called for four more reservoirs, including Marvin Nichols Reservoir in neighboring Region D, which would mainly supply Region C. Region D refused to put the reservoir in its plan. In 2007 the Texas legislature passed a bill designating the Marvin Nichols Reservoir in the State Water Plan. The bill also created a study commission to resolve the differences between Regions D and C. Groundwater use in the region has actually decreased since 1980, and there are no significant springs in the

The Trinity River in downtown Dallas flooded in March 2006. Photo courtesy of the Trinity River Authority.

Dallas–Fort Worth area used for water supply. Many springs in the area have dried up or have reduced flows due to historical overpumping of groundwater.

More than half of the municipal use of the area is returned to the streams after treatment in numerous wastewater plants. There are several projects under consideration for reusing this wastewater that will require permits from the state. The largest wastewater treatment plants in Region C discharge into the Trinity River and its tributaries downstream from the reservoirs. In summer the flows downstream are actually higher than under natural conditions, the result of reservoir management, wastewater discharges, and flows imported from other watersheds.

For more than one hundred fifty years, the idea of navigating via the Trinity River from the coast to Dallas has been a dream. Since the 1830s steamboats have plied the Trinity to varying degrees, with one boat getting within 40 miles of Dallas in 1854. Snags, sandbars, and low water made navigation unpredictable, and the Civil War halted any major improvements to the waterway. Nevertheless, in 1868 a cargo boat finally reached Dallas after a one-year voyage. In spite of the growth of the railroads, several Trinity navigation projects were considered, and some dams and locks were actually constructed in the early 1900s. The project was abandoned with the onset of World War I. In the 1950s and 1960s various storage and flood control reservoir projects were designed to facilitate possible future barge traffic. In 1973 voters were asked to fund a $1 billion navigation project on the Trinity, but the measure failed. Navigation of the Trinity still remains a dream for some, though it seems unlikely to become a reality.

Currently, broader awareness of the benefits of the Trinity River to the Dallas–Fort Worth area has resulted in intercommunity plans for restoration of the riverine character of the four forks that constitute the Upper Trinity basin. In partnership with the National Park Service, the Army Corps of Engineers, and the Meadows Foundation and other private philanthropies, both Dallas and Fort Worth have begun major projects that include river restoration, river trail networks, cleanups of riverside landfills, and establishment of nature centers to educate the public about the river's natural history and benefits to society.

The Sabine and Neches Rivers of East Texas

THE SABINE, TEXAS' EASTERN BOUNDARY Like the other borders of Texas, the eastern boundary consisting mainly of the Sabine River was altered and disputed. Texas and Louisiana were controlled by Spain, France, England, and, finally, the United States. Until an 1819 treaty between Spain and the United States, the eastern border of Texas was considered to be the Arroyo Hondo. This tributary of the Red River lies in what is now Louisiana and was used by the French and Spanish as the Texas-Louisiana border in the 1700s. After the Sabine River was established as the Texas border, arguments continued for more than one hundred fifty years over whether the border was the middle of the stream or the bank. In 1976 the U.S. Supreme Court settled the matter: it declared the middle of the river as the boundary. Oil leases worth several million dollars on the disputed land could have played a part in the two states' perseverance.

First light on the Neches River. Photo by Adrian Van Dellen.

The Sabine originates in northeast Texas and only becomes the border at the thirty-second parallel. Unlike most Texas rivers, the Sabine Basin receives abundant rainfall, and the Sabine River discharges the largest volume of water of all Texas rivers. Rainfall averages in the Sabine Basin range from 37 to 50 inches. The Sabine and Neches Rivers, which converge near the coast, together discharge the amount of water equal to fifty times the volume of their estuary. This huge flow dwarfs that of the Trinity and Guadalupe Rivers, which discharge four to seven times the volume, and the Nueces River, which discharges less than 60 to 70 percent of its estuary volume.

The Caddo Indians reached the area in A.D. 780 and were the southwesternmost example of Mississippi mound-building culture. European settlers encountered Caddos living along the Sabine when they came to the area in the 1500s. The word *sabine* is Spanish for "cypress," the large trees that grow along the edge of the lower river.

During the years of the Texas Republic and into the mid- and late 1800s, lumber and cotton were transported by steamship on the Sabine River. This led to the rise of Port Arthur and Orange as ports for the Sabine and the Neches Rivers, which converge at an inland estuary before reaching the Gulf of Mexico. Timber continues to be the primary natural resource of the Sabine-Neches Basin. By 1900 oil was discovered at Spindletop oilfield near Beaumont, marking the birth of the modern petroleum industry. Beaumont, with a population of 113,866 in 2000, is the largest city in the Sabine-Neches Basin. The ensuing development of oil refineries, shipping facilities, and wastewater discharges resulted in the deterioration of the water quality of the Lower Sabine River, which, although being addressed, still remains a problem.

About 80 miles northeast of Beaumont the Sabine River is impounded in Toledo Bend Reservoir, stretching for more than 100 miles along the border between Louisiana and Texas, which share the water and power generated by the dam. Toledo Bend, covering 181,600 acres and with storage capacity of 4,472,900 acre-feet, is the fifth largest man-made body of water in surface acres in the United States. The reservoir was built primarily for flood control and hydroelectric power but is destined to become increasingly vital as a source of water from growing urban areas farther west.

THE NECHES RIVER The Neches River drains the area just west of the Sabine Basin and shares many of the same characteristics. The Caddo Indians were also some of its early inhabitants. Steamboats were used in its lower stretches to transport cotton to the ports around Sabine Lake where the Neches joins the Sabine River. Both basins have an abundance of pine and hardwood forests, and lumber and forest products are a major part of the basin economy. The Neches Basin, like the Sabine, remains relatively uninhabited; Beaumont is its only major city.

Wildlife habitat and recreation areas are in abundance in the two watersheds and include:

- Eight state wildlife management areas
- Ten state parks
- Three federal wildlife refuges, including the new Neches River National Wildlife Refuge created in 2006
- Four state forests
- Two Corps of Engineers lakes
- Three national forests

Like the Sabine Basin, the Neches Basin has high average rainfalls for Texas, and its average annual flow is about 6 million acre-feet per year. A major reservoir, Sam Rayburn Reservoir, on the Angelina River (a tributary of the Neches), is the largest lake completely contained in Texas. With a capacity of 3,997,600 acre-feet, Sam Rayburn covers 114,500 acres and is considered one of the top bass fishing lakes in the nation. Unfortunately, a fish consumption advisory is posted for the reservoir due to mercury contamination. In addition, portions of the reservoir, in particular the upper end, have historically had problems with metals contamination, pH, dissolved oxygen, and nutrients. Paper mills and wastewater discharges are some of the causes of these problems.

Estuarine inflows from the Neches have been a concern for a number of years. Construction of Lake Palestine in its upper reaches during the early 1960s was contingent on mandating a 5 cubic feet per second (cfs) release, one of the first such requirements in the state. In 2003 a saltwater barrier was completed on the Neches at Beaumont. The barrier, with adjustable gates, prevents salt water from migrating up the river channel at low flows so as not to contaminate upstream intake pipes. The effects of this structure can be seen even as far upstream as Sam Rayburn Reservoir, where freshwater historically has had to be released to push the saltwater downstream—an activity that caused a great deal of controversy among lakeside landowners.

Although the Sabine and Neches Basins have two large reservoirs and plentiful rainfall, groundwater supplies approximately 25 percent of the total water consumed in the water planning region in which they are located. The two main aquifers in the area are the Gulf Coast and the Carrizo-Wilcox. Both aquifers extend northeast to southwest across the coastal plains of Texas. More than 250 springs have been documented in the planning region, but as most of these have flows of less than 10 gallons per minute, they are not considered significant water sources.

THE RIVERS, AQUIFERS, AND SPRINGS OF THE SOUTH-CENTRAL TEXAS REGION

The Hill Country of Texas comprises generally rocky limestone hills covered with oaks, mesquite, and the ubiquitous cedar

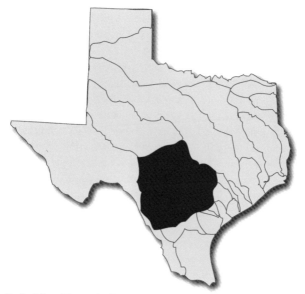

The South-Central Texas region. Data Source: Texas Parks and Wildlife Department and Texas Water Development Board.

breaks that are actually Ashe junipers. This area stretches along the edge of the Edwards Plateau near Waco, south through Austin and San Antonio, then west to Del Rio, with rainfall averages from 22 inches in the west to 34 inches at the eastern edge. The Colorado River starts in far northwest Texas and cuts its way through the Hill Country. Other rivers of south-central Texas either originate on the Edwards Plateau—the Llano, Guadalupe, Nueces, and Frio Rivers—or start at the base of the Balcones Fault—the San Marcos, Comal, and main stem of the San Antonio Rivers.

The Edwards Plateau and Its Aquifers

The Edwards Plateau comprises almost 31,000 square miles of south-central Texas. It is bounded on the west by the Pecos River, on the east and south by the Balcones Escarpment, and on the north by the plains and the granite of the Llano Uplift.

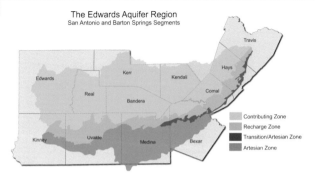

The Edwards Aquifer Balcones Fault Zone. Data Source: Gregg Eckhardt and www.edwardsaquifer.net.

Underlying the plateau is the Edwards Formation, consisting mainly of limestone formed from coral and shells from ancient seas. As the seas receded, the Edwards Formation was eroded and later covered with sediment. Cavities created by erosion produced pockets that could hold water. These cavities are today's aquifers. Faults, layers of various materials, and other geologic features have separated areas of the formation, resulting in several aquifers that underlie the Edwards Plateau today. Some of these aquifers are isolated; some are connected to varying degrees with each other.

The Edwards-Trinity (Plateau) Aquifer extends from the Hill Country of Central Texas past the Pecos River into the Trans-Pecos region. The Edwards-Trinity (Plateau) Aquifer includes areas under the Stockton Plateau, which is west of the Pecos River and is sometimes referred to as the Western Edwards Plateau. The Edwards-Trinity (Plateau) Aquifer is considered a major aquifer. To make matters more confusing, there is a minor aquifer to the north called the Edwards-Trinity (High Plains) Aquifer. This aquifer underlies parts of the Ogallala Aquifer in Texas.

When people in Texas refer to the Edwards Aquifer they are usually talking about the Edwards (Balcones Fault Zone, or BFZ) Aquifer. The Edwards (BFZ) extends from an underground divide in Kinney County in the west through San Antonio and northeastward to Bell County. This relatively narrow aquifer has three distinct sections separated by groundwater divides. The San Antonio segment is the largest and stretches from Kinney County almost to Kyle. This segment ranges from 5 to 40 miles wide and is 175 miles long. From Kyle to the Colorado River in Austin is the Barton Springs segment. The northern Edwards Aquifer stretches from the Colorado River north to Bell County. The groundwater divides keep the waters of these three segments from mixing.

The porous limestone of the Edwards Formation permits water to enter and move through it rapidly. This characteristic allows for quick recharge but also reduces its storage capacity and limits its ability to filter out contaminants in runoff flowing into the aquifer. Flow rates in the Edwards (BFZ) range from a few feet to 1,000 feet per day. The average residence time for the Edwards (BFZ) water is about two hundred years, a young age when compared to the water in the Ogallala Aquifer, which does not recharge easily and can be ten thousand years old.

MAJOR SPRINGS OF THE EDWARDS FORMATION All of the largest springs in Texas flow from the Edwards (BFZ) or associated Edwards limestone formations to the west. The largest spring in the United States west of the Mississippi River is Comal Springs in the city of New Braunfels. With an average flow of 284 cubic feet per second (cfs), the Comal surpasses the second-ranked San Marcos Springs, which averages 172 cfs. Interestingly, especially to hydrogeologists and water managers, Comal Springs went dry during the drought of record in the 1950s, while the San Marcos continued to flow at 46 cfs.

The Comal and San Marcos Springs lie only 16 miles apart at the eastern edge of the Edwards (BFZ) Aquifer along the Balcones Escarpment. Each spring group, consisting of several smaller springs at the headwaters, provides the flow for the Comal and San Marcos Rivers respectively. The year-round 72 degree F temperature of the two rivers and the fairly steady rate of flow create habitat for several endangered species. Numerous rare and endemic invertebrates found in Hill Country caves are also dependent on the cool temperatures, high humidity, and food supply afforded by flows through the aquifer.

The endangered fountain darter was native to both springs but could not be found in the Comal River after the drought of the 1950s. In the 1970s the fountain darter was successfully reintroduced into the Comal from a population in the San Marcos River where it still survives. The Comal is home to two endangered aquatic beetles and an endangered amphipod. The San Marcos River is also the habitat for the endangered San Marcos cambusia, which has not been sighted since 1983. Endangered Texas wild rice grows in the San Marcos River and the Texas blind salamander is found in the Edwards Aquifer near the springs. In all, there are eight endangered or threatened species in the two spring systems, the nearby rivers, and the Edwards Aquifer.

To protect the springs as habitat for these endangered species, the Sierra Club and other entities, including the Guadalupe Blanco River Authority, filed a federal lawsuit under the Endangered Species Act in the early 1990s. The result was the establishment of the Edwards Aquifer Authority (EAA), which regulates pumping of the aquifer. Groundwater in Texas is generally unregulated, and this was even more true in the early 1990s.

The 2007 Texas legislature significantly altered management of the San Antonio portion of the Edwards (BFZ) by increasing the permitted pumping to 572,000 acre-feet. The original legislation adopted in 1993 allowed only 450,000 acre-feet, which was supposed to decrease to 400,000 in 2008. The new law does take steps to reduce the impact of the increased pumping during drought; pumpers must now reduce their pumping more quickly than was required under the old rules.

The legislation also includes a Recovery Implementation Program (RIP) for the threatened and endangered species found in the aquifer and springs. A steering committee will appoint a science subcommittee to develop recommendations for withdrawal reductions and stages of critical management during droughts. By September 2012 the steering committee, with

The Texas Rivers Center and glass bottom boat in San Marcos. Photo by Jason Taylor.

input from stakeholders and the science subcommittee, will make Edwards (BFZ) management recommendations to the Edwards Aquifer Authority.

The characteristics of constant flow and temperature have historically attracted humans to the San Marcos Springs. Excavations at the bottom of Spring Lake, which was dammed in 1848, reveal that humans have lived near the springs for twelve thousand years, making it one of the oldest continuously inhabited sites in North America.

Today Texas State University–San Marcos owns the land around the San Marcos Springs at the headwaters of the San Marcos River. In cooperation with the Texas Parks and Wildlife Department, the university established the Texas Rivers Center to serve as an educational and research facility focusing on the importance of aquifers, rivers, and aquatic systems in Texas. The center houses the Rivers System Institute of Texas State University, Texas Parks and Wildlife Department's Freshwater Resources Program and Contaminants Assessment Team, and

Barton Springs is a popular swimming pool in downtown Austin, Texas. Photo by Angeline M. Van Zant.

the National Park Service's River Trails and Conservation Assistance Program for Texas.

Just to the north of San Marcos, near Kyle, is the Barton Springs Edwards Aquifer, which is isolated hydrologically from the Edwards (BFZ). Although the Barton Springs Edwards is a minor aquifer, it faces major impact from development. The results are most noticeable at Barton Springs, the fifth largest springs in Texas, which feed a historic swimming area in Austin. Water quality deterioration and reduced springflows have prompted the city of Austin to implement landmark land use regulations to protect the recharge area of Barton Springs. In 1997, as a result of a lawsuit filed by the Save Our Springs Alliance, the Barton Springs salamander was declared an endangered species. This salamander lives at or near the mouth of Barton Springs and is similar to the Texas blind salamander. They are adapted to an entirely spring-fed aquatic life in the darkness of limestone cavities and vegetation cover.

To the west of the Edwards (BFZ) are the third and fourth

largest springs in Texas. Goodenough Springs still flows from beneath the Amistad Reservoir, and San Felipe Springs provides both water supply and recreation for the city of Del Rio. Both springs flow from the Edwards Formation near the Rio Grande.

THE TRINITY AQUIFER The Trinity Aquifer stretches from Central Texas to Oklahoma; however, the Trinity is made up of several aquifers. The Travis Peak Formation of the Trinity underlies portions of the Hill Country. Geologists agree there is interchange between the Edwards (BFZ) and the Trinity but differ on the amounts: from around 50,000 acre-feet per year to over 300,000. Unlike the Edwards (BFZ), the Trinity is slow to recharge.

Portions of the Trinity Aquifer underlie the Edwards (BFZ). Some wells are drilled through the highly regulated Edwards (BFZ) to underlying portions of the Trinity Aquifer, which often are in unregulated groundwater areas. The Trinity Aquifer is subject to overpumping in many places in the Hill Country west of Austin, which is of special concern because of the Trinity's sluggish recharge.

Jacob's Well, a popular spring originating from the Trinity Aquifer and the heart and soul of Wimberley, provides summertime base flow for the Blanco River, a tributary of the San Marcos and Guadalupe Rivers. There was no record of Jacob's Well going dry until 2000. The slow recharge of this portion of the Trinity and increased pumping for the fast-growing Central Texas region are blamed for the flow stoppage. The 2000 drought lasted only a matter of months—a short time for Texas droughts. Jacob's Well resumed flowing after above-average rainfall years but is predicted to go dry in another similar drought. As an example of the complex hydrogeologic relationship between the Edwards and the Trinity, after Jacob's Well flows a few miles and joins the Blanco River, a portion of the Blanco flow is recharged into the Edwards (BFZ) Aquifer a few miles farther downstream, illustrating how the overpumping of one aquifer can affect another.

A famous commercial in the 1970s promoting Texas-brewed

beer proclaimed it was "from the country of 1100 springs," in reference to the Hill Country and its numerous springs. Whether or not 1,100 was ever the actual number of existing springs, there are certainly fewer springs now as a result of overpumping.

The Colorado River and the Highland Lakes

The Colorado River, the longest river entirely within the boundaries of Texas, or any single state, carves its way through the rocky, flood-prone Hill Country, forming impressive canyons. These canyons proved efficient locations for reservoirs to control flooding, generate power, and provide water for a growing Austin and Central Texas.

There are six Highland Lake dams on the Colorado in Central Texas. The first one was built in the 1890s near Austin but was destroyed on more than one occasion by floods and overflowed in 1935. The other reservoirs were subsequently built upstream of Lake Austin. President Lyndon B. Johnson, a native of Central Texas, was instrumental in the construction of many of these Highland Lake dams, which electrified the impoverished Hill Country, tamed the most damaging floods, and launched his political career. The Lower Colorado River Authority (LCRA), created in 1934, is the organization responsible for the management of the Colorado River from the Highland Lakes to the coast.

Lake Buchanan is the uppermost Highland Lake and the largest in surface area: 22,335 acres. Completed in 1937, the two-mile-long Buchanan Dam is the longest multiple-arch dam in the United States. This form of construction is no longer used because it is so labor intensive; however, modern dam construction can use up to one thousand times more materials. Upstream of Lake Buchanan where the river still flows in its natural state lies Colorado Bend State Park, an impressive canyon with springfed waterfalls dropping into the Colorado River. Birdwatchers enjoy the bald eagles that winter in this canyon, and the spring run of white bass is one of the state's premier fishing experiences.

The Colorado River's Highland Lakes and the dams that created them. Data Source: Lower Colorado River Authority.

Downstream of Lake Buchanan, two tributaries of the Colorado that arise in the Edwards Plateau flow into the Highland Lake chain. The Llano River, which drains over 4,000 square miles, enters Lake LBJ from the west. A smaller river, the Pedernales, enters Lake Travis from the west. *Llano* is the Spanish word for "plains"; *Pedernales* is Spanish for the type of flint rocks in the riverbed. Both the Llano and Pedernales are flood-prone rivers. In 1935 the Llano peaked at 380,000 cfs at the city of Llano upstream of Lake LBJ. The Pedernales River, with only 900 square miles of drainage, reached 441,000 cfs at Johnson City in 1952. In comparison, the largest recorded flow at Austin was 555,000 cfs in 1869 from a drainage of 39,000 square miles. The flood potential of these small and narrow limestone channel rivers of the Edwards Plateau when combined with the rainfall potential of Central Texas is extraordinary and at times has been devastating.

Lake Travis, just north of Austin, has the largest holding capacity of the Highland Lakes. It is also the only one in the chain designed specifically for flood control with a storage capacity of 369 billion gallons. Several of the Highland Lake dams generate electricity, including the dam at Lake Travis.

Together, the Highland Lakes are used to manage floods and generate power. LCRA also uses the lakes to control the entire Lower Colorado River all the way to the coast. Large quantities of water are used for rice farming in Colorado, Austin, Wharton, and Matagorda Counties near the coast. The LCRA manages the Highland Lakes to provide steady flows at the proper times in these areas. The overall LCRA management plan for the Lower Colorado includes prescribed releases for freshwater to maintain the Matagorda Bay and Estuary.

Who owns water in Texas is a contentious issue. For example, the city of Austin and the LCRA have historically disagreed over which entity controls the millions of gallons of treated wastewater, or return flows, discharged from the city. In 2005 both entities applied for the rights to these return flows, and each filed a protest against the other's permit application in a historic struggle over wastewater unthinkable a generation ago. In 2007 the two parties agreed to share the rights to the wastewater.

The Guadalupe—Central Texas Water Park

The Guadalupe River, one of the most popular recreational rivers in Texas, emerges from springs in the Edwards Plateau in Kerr County and flows for about 90 miles before being impounded in Canyon Lake. The 16-mile stretch below Canyon Dam to New Braunfels is the tubing capital of Texas and home to highly successful rainbow trout fishing. On summer weekends, thousands of people descend on this stretch of river for recreation, mainly inner tubing.

Canyon Lake, the only major reservoir in the Guadalupe Basin, is under tremendous pressure as it faces multiple and competing demands. San Antonio needs more water; downstream recreational interests want flows released in the summer; environmentalists demand more environmental flow for the Lower Guadalupe, particularly for varying seasonal needs in the estuary; and lakeside landowners want a full lake that comes up to their boat docks. Another interest group, Trout Unlimited, won an out-of-court settlement in 2001 for increased releases from the lake to keep the stocked trout alive over the summer for the next season. Several hydropower facilities owned by the Guadalupe Blanco River Authority (GBRA), including Canyon Dam, also need flows for generating electrical power. River authorities are quasi-governmental agencies that manage and develop water resources in their respective river basins. (See chapter 5 for more on river authorities.)

The GBRA manages Canyon Lake along with the U.S. Army Corps of Engineers. Many lakes in Texas are jointly managed through such partnerships between the Corps of Engineers or the Bureau of Reclamation and a local or regional entity. The Highland Lakes are an exception in that the LCRA has exclusive control over the reservoirs and dams, undoubtedly the result of its unique political history. Under a shared control agreement reservoir like Canyon Lake, the flood storage portion of the reservoir, or flood pool, is controlled by the Corps. The other

Tubers enjoy the water of the Guadalupe River. Photo courtesy of Texas Parks and Wildlife Department.

After intense rainfalls, Canyon Lake overflowed its spillway in 2002. Photo by Tom Hornseth and courtesy of the Comal County Engineers Office.

managing entity, a river authority for example, will control the conservation pool, which is water stored for uses such as municipal supply, hydroelectric power, and recreation.

The Guadalupe River, like most Hill Country rivers, is prone to flash floods. In 2002 Canyon Lake overflowed its spillway for the first time since its construction in 1966. The path of the spillway water cut a new, deep channel through areas where residents, not anticipating a dam overflow of this magnitude, had built homes. In the late 1990s and early 2000s several major floods ruined homes in and around New Braunfels and downstream. This pattern is expected to intensify as urban development and land fragmentation continue in the watershed.

At the end of the 16 miles of rapids and pools below Canyon Dam lies New Braunfels, a growing Central Texas community whose population is about forty thousand. Here the Guadalupe is joined by the Comal River, which, at 3 miles, is said to be the shortest river in the United States. At its headwaters are the Comal Springs. During periods of low flows on the Guadalupe

River, the springs can contribute much of the base flow of the lower river all the way to the estuary.

New Braunfels is also well known for being one of the early German settlements in Central Texas and still retains much of its German heritage, including the annual Wurstfest, a popular celebration of German culture. Beginning in the 1830s, German entrepreneurs led several thousand Germans to settle in Texas, many of whom made their way to the Hill Country. They were successful farmers and built mills to process their grain and corn. Many of the Hill Country rivers and streams still have remains of these mills and their associated dams. These rivers were ideal for this purpose because the springs produced a dependable flow of water.

Downstream of New Braunfels, the San Marcos River, with headwaters at San Marcos Springs, enters the Guadalupe River. San Marcos Springs did not dry up in the 1950s drought and continues to provide a substantial portion of the flow of the Guadalupe River in drought periods. The San Marcos River begins in the city of San Marcos, where the springs flow from the bottom of Spring Lake. A few miles downstream of San Marcos, the Blanco River, with a drainage of over 400 square miles, enters the San Marcos River on its way to the Guadalupe—hence the name of the managing entity, the Guadalupe Blanco River Authority. Although the Blanco ceases to flow in dry periods, it is another flood-prone Hill Country river with a record flood of 139,000 cfs.

The San Antonio River

The San Antonio River historically originated from headwater springs on the campus of the University of the Incarnate Word that now only occasionally flow when the Edwards Aquifer reaches historic heights after heavy rains. Most of the flow now comes from reused municipal wastewater that is recirculated through a complex system of canals and tunnels. From this murky beginning comes one of the world's most famous stretches of urban waterway.

The San Antonio Riverwalk winds for a few miles through

several blocks of downtown San Antonio and is a mecca for tourists and conventioneers from around the world. Restaurants, bars, and clubs sit only a few feet above the water line. Decorated barges carry tourists along the riverway day and night, dense plantings of tropical plants soften the landscape, and arched bridges link the shores of this riverine fantasyland. On a limestone plaza high above the riverwalk sits the most visited tourist destination in Texas—the Alamo, named in Spanish for the cottonwood trees that once lined the riverbank.

The history of San Antonio is probably the richest of any city in Texas, and the Alamo is the icon of Anglo Texas history. Here, in a mission established by the Spanish in 1724, 189 Texas volunteers held off thousands of Mexican soldiers for thirteen days. The Mexican Army prevailed, and all the Texans were killed. However, the courage of the Texan garrison became a rallying cry for the Texas Revolution, and not long afterward Texas defeated the Mexican Army at the Battle of San Jacinto. The battle cry, "Remember the Alamo," has become famous the world over.

The San Antonio Riverwalk in spring. Photo by Gregg Eckhardt and courtesy of www.edwardsaquifer.net.

The Spanish had the greatest cultural influence on San Antonio. Since Spain had an arid climate like Texas, the Spanish came prepared for surviving in a dry climate. They diverted the San Antonio River into ditches, or acequias, to distribute water throughout the area. Their management of water was so precise that they could separate plots of land into 225 classifications according to soil fertility, slope, and suitability for crops. The 250-year-old Espada Acequia is the most complete surviving Spanish irrigation system in the United States.

San Antonio's municipal water uses have also helped to shape the city. Patterns of growth and economic divisions can be traced to Colonel Brackenridge's private water infrastructure development and subsequent Belgian investors who extended infrastructure on a for-pay basis.

As early as 1887 city planners discussed the idea of creating a San Antonio River park. Since then, millions of dollars have been spent to create a river walk, on infrastructure to prevent flooding in the downtown area, and to recirculate the water itself. Even though the river's origin is derived from several springs that no longer flow, there is an upstream watershed subject to the Hill Country deluges. A flood diversion tunnel completed in 1996 is 3 miles long and 24 feet in diameter and lies 150 feet underground. It is capable of carrying 6,700 cfs of floodwater. Under normal conditions, this tunnel is used to recirculate water for the San Antonio Riverwalk, so that water for riverflow now seldom needs to be pumped from the Edwards Aquifer.

The Medina River, a tributary of the San Antonio, arises from springs in the Edwards Plateau northwest of the city. Although not considered a major Texas river, the Medina provides recharge for the Edwards Aquifer and is used for irrigation. Medina Lake, on the Upper Medina River, was completed in 1912 and financed by the British. At that time, the lake had the fourth largest dam in the United States with a capacity of 254,000 acre-feet. In 2002 record flood flows came within a foot of topping out the dam, and there was fear that the dam would fail. Residents living downstream were ordered to evacuate, but the dam held. Discussions about its integrity continue. The Me-

The Landmark Inn State Historic Site is on the banks of the Medina and the Old San Antonio Road. Photo courtesy of Texas Parks and Wildlife Department.

dina is also the site of San Antonio's first proposed municipal supply lake, the Applewhite Reservoir, which has been repeatedly voted down by city residents over environmental, financial, and equity concerns. This failure to develop additional surface water supplies has resulted in ever greater reliance by the city on Edwards (BFZ) Aquifer water.

As the Medina leaves the Hill Country, it passes through historic Castroville. Here in the mid-1800s settlers from the Alsace region of France and Germany established the community. A dam and major mill on the river are part of a historic park in the town, the Landmark Inn State Historic Site. The inn was used primarily as the only place to take a bath between San Antonio and the border. The Medina River continues through the South Texas brush country and meets the San Antonio River three counties below San Antonio. The San Antonio River meets the Guadalupe River only about 10 miles before the coast; legally, however, they are defined as different river basins—a situation of constant controversy that complicates management of water resources in the region.

The Nueces River

The Nueces River drains some of the more rugged and arid parts of Texas and is the only major drainage between the San Antonio River and the Rio Grande. The basin remains a sparsely inhabited area of the state; the only major city is Corpus Christi, located at the mouth of the river. Fed by springs from the Edwards Plateau, the Nueces' 16,200-square-mile drainage averages only 620,000 acre-feet a year in runoff. The headwaters watershed is used mainly for sheep and goat ranching and is considered excellent deer hunting country. Near the headwaters of the Nueces, the smaller Frio and Sabinal Rivers also start their spring-fed journey, eventually joining the Nueces after passing through the Balcones Escarpment. Garner State Park, one of the most visited state parks in Texas, lies on the banks of the Frio—a mecca for tubers and paddlers seeking respite from the hot South Texas summers.

Although never a center of Mexican or Texan culture, the

Old Baldy at Garner State Park is beautiful in the fall. Photo courtesy of Texas Parks and Wildlife Department.

Nueces was the first river to appear on European maps. On a 1527 Spanish map, the mouth of the Nueces is shown behind its barrier island. It was not until 1747 that the Upper Nueces was linked with the mouth of the river—testimony to the remoteness of the area. For two hundred years explorers thought the Nueces was a tributary of the Rio Grande. After the discovery of its connection to what is now Nueces Bay, the river became a disputed boundary between Texas and Mexico. The area between the Nueces and the Rio Grande remained disputed territory from the Texas Revolution until the end of the Mexican War.

The Edwards Plateau region of the Nueces and Frio Rivers has no significant water development projects. However, the two rivers contribute a significant amount of recharge to the Balcones Fault Zone portion of the Edwards Aquifer. Downstream, two major reservoirs, Choke Canyon and Lake Corpus Christi, store water for the growing Corpus Christi region. Upstream, these two rivers run wild and free, with flows varying from practically nothing to raging torrents. These are two of only a few remaining free-run stretches of rivers in Texas.

THE CENTRAL AND LOWER RIO GRANDE REGION
The Central Rio Grande

"Central Rio Grande" generally refers to the portion of the Rio Grande where the cities of Del Rio, Eagle Pass, and Laredo are situated. Below Lake Amistad at Del Rio, there are no significant tributaries of the Rio Grande until the Rio Salado enters Falcon Reservoir from Mexico more than 200 miles downstream. This is generally a sparsely inhabited stretch of the river, and it remains so downstream to Laredo. The two major Texas reservoirs on the Rio Grande, Lake Amistad and its partner, Falcon Reservoir, operate as a system to control the Rio Grande all the way to the coast.

Swimming is not advised in the river below Lake Amistad between Del Rio and the confluence of Rio Salado in Val Verde County, due to high fecal coliform counts. As in several other parts of the Rio Grande, this contamination is the result of a combination of little or no wastewater treatment in border cit-

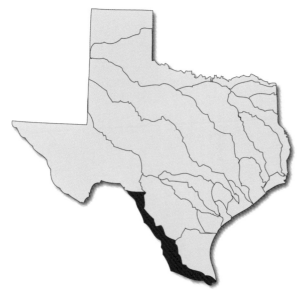

The Central and Lower Rio Grande region. Data Source: Texas Parks and Wildlife Department and Texas Water Development Board.

ies in Mexico, poor or nonexistent septic tanks on the U.S. side, and little or no wastewater treatment in largely ungoverned communities, or *colonias*, in the United States.

As early as 1665, a Spanish army contingent crossed the Rio Grande on what is known as San Antonio Crossing near present-day Eagle Pass. This crossing was heavily used since it was on the route of the Camino Real de los Tejas connecting San Antonio with Mexico City. Laredo, and its border counterpart Nuevo Laredo, is the largest border crossing in the United States for goods going to and from Mexico.

Since the passage of the North American Free Trade Agreement (NAFTA) in 1993, traffic between Mexico and the United States has increased significantly. The volume of trade through Los Dos Laredos (The Two Laredos) had increased to 88 percent of all U.S.-Mexico trade by 1997. From 1992 to 1997 the number of U.S.-owned manufacturing facilities (maquiladoras) in border towns increased from 190 to 340. In 1997 more than 1.2

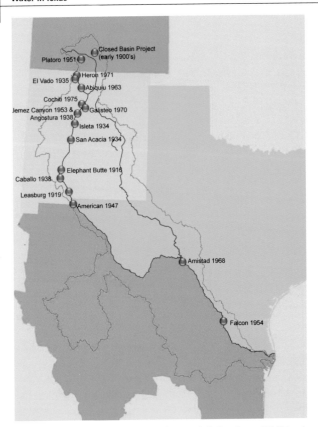

The dams along the Rio Grande and the dates they were built. Data Source: U.S. Fish and Wildlife Service.

million trucks and 245,000 railcars passed through Los Dos Laredos. The resulting economic boom has enabled Nuevo Laredo to expand and upgrade its wastewater systems; however, it has also created new primitive settlements, called *colonias*, near the town that lack sufficient infrastructure. Thus water quality on the Rio Grande remains a problem in the Laredo area.

Lake Amistad and Falcon Reservoir

Falcon Reservoir downstream of Laredo was built in 1954. Its partner reservoir, Lake Amistad, was finished in 1968 and was designed to work in conjunction with Falcon. Together they have made possible almost complete control of any flooding that might originate in the area above Falcon Dam, which includes the Rio Salado from Mexico. The two lakes combined provide storage for over 8 million acre-feet of water for the two countries, including 6.05 million acre-feet for water supply and 2.25 million for flood control. The United States funded and controls 56.2 percent of Amistad and 58.6 percent of Falcon Reservoir; Mexico controls the rest. The projects are operated and maintained by the International Boundary and Water Commission.

Since Amistad is upstream of Falcon and receives about twice the average inflows than the latter, the sedimentation rate is much greater in Amistad. Projections for reduction of storage space due to sediment buildup from 2000 to 2050 show a loss of 392,037 acre-feet, or 23 percent, of the storage capacity—just for the U.S. portion of Amistad. Falcon Lake has negligible projected losses from sedimentation. Loss of reservoir capacity due to sedimentation will be a major cause of water supply reductions in the future. Measures will have to be taken to make up for this decrease.

Watershed management in Mexico also affects both lakes. In addition to the Rio Conchos flowing into Amistad and the Rio Salado flowing into Falcon, the Rio San Juan enters the Rio Grande below Falcon Reservoir. The combined reservoir storage of the tributaries in Mexico is 6,240,000 acre-feet. The Mexican tributaries above Falcon Dam hold 4,410,000 acre-feet, while the Rio San Juan reservoirs hold 1,833,000 acre-feet.

In the 1944 treaty the obligations of Mexico and Texas for allowing water to leave their respective countries were defined. Mexico is required to send the United States an average of 350,000 acre-feet annually from its six tributaries of the Rio Grande. In return, the United States must send Mexico 1.5 million acre-feet from the Colorado River, which flows through Utah, Nevada, and Arizona to the Gulf of California. In the water plan for Region M, including the Lower Rio Grande Valley, the flow models assume that in a drought of record there will be no water supplied from Mexico. Although the 1950s are still the drought of record for the

U.S. side, the water plan states that the 1990s drought may prove to be the record drought for the Mexican contributing basins—a complex challenge for water managers.

In 2005 Mexico began paying off a water debt of 733,000 acre-feet that had been steadily increasing for twelve years. By 2002 the lack of water in the Lower Rio Grande was causing U.S. farmers to go out of business. The Rio Grande had actually stopped flowing to the sea at various times between 2000 and 2002. The debt became as high as 1.5 million acre-feet until heavy rains in 2003 and 2004 replenished much of the two Rio Grande reservoirs. Although groundwater is available from the Carrizo-Wilcox and Gulf Coast Aquifers, there has not been significant usage in the Lower Rio Grande area. There are no major springs south of the San Felipe Springs in Del Rio that flow from the Edwards Plateau.

A complex computer model is used to manage Lake Amistad and Falcon Reservoir and the irrigation and municipal needs downstream. Irrigation system loss and river channel loss due to evaporation and seepage below Falcon Lake is estimated to be between 29 and 52 percent. These losses require that extra water be released over and above the requested amount to arrive downstream at the use point. A significant factor in the management of these releases is what is termed irrigation carrying water. If a municipal demand is released at the same time as an irrigation demand, then there are less overall seepage and evaporation losses. However, drought management of the Amistad-Falcon system calls for reduced irrigation releases in low-water conditions in favor of municipal releases. If municipal water is released without irrigation carrying water, then there will be a higher percentage of evaporative loss due to a smaller slug of water moving downstream more slowly.

The Lower Rio Grande

Approximately 80 percent of the water use in the Lower Rio Grande Valley is for irrigation. The fifty-year 2007 State Water Plan projects that this number will decrease from 1.16 million to 981,748 acre-feet. The estimated decrease is based on irrigated

The Lower Rio Grande Valley citrus crop has an estimated economic impact of more than $200 million annually. Photo from iStock.com.

acreage being converted to urban areas. This region is one of the fastest growing in the United States—from a population of 400,000 in 1950 to 1.7 million in 1998 and a projected 3.8 million in 2060. Most of this growth is expected to occur in Laredo and the Lower Rio Grande Valley. Added to these growth projections is the Mexico portion of the Rio Grande Basin. The Region M water plan estimates that almost 7 million people currently live in the Mexico portion of the Rio Grande Basin.

The Lower Rio Grande Valley, known as the Valley, encompasses Starr, Cameron, Hidalgo, and Willacy Counties and is actually the delta of the Rio Grande. As a delta, this area has alluvial soils ranging from sandy loam to clay and, combined with its southern latitude equivalent to South Florida, makes for an excellent agricultural area, specifically, for citrus fruit. The Spanish planted orange trees in the 1700s north of Edinburg, Texas. Commercial citrus production did not take off, however, until the railroads came in 1904. Grapefruit became the major crop, with over 7 million trees in 1932. Periodic freezes can kill off half the trees, but the industry continues to flourish and ranks third in U.S. citrus production behind Florida and California-Arizona.

For its last 180 miles, the Rio Grande is a series of small in-

stream dams, levees, flood control channels, irrigation channels, and the Arroyo Colorado. The Arroyo Colorado is an old
channel of the Rio Grande that flows into the Laguna Madre,
the southernmost estuary of Texas. The arroyo serves many
uses: irrigation water is delivered, surplus irrigation water is
carried off, municipal wastewater is discharged, floodwaters
are diverted, fishing is allowed, and barges travel from the Intracoastal Waterway to Harlingen. Predictably there are water
quality problems in the arroyo such as depleted oxygen and
toxic chemicals. Between 1990 and 2004, 26 million fish died in
19 fish kills documented by the state. Since 1998 the Texas Commission on Environmental Quality has conducted programs to
clean and restore the Arroyo Colorado. Despite these problems,
the Arroyo Colorado is still the major source of freshwater to
the Lower Laguna Madre.

Starting 180 miles upstream of the mouth of the Rio Grande,
a series of levees on both sides of the river control flooding and
divert higher flows into the United States and Mexico through
complex agreements and engineering. Two small instream dams
facilitate these diversions: the Anzalduas and Retamal Diversion
Dams. These two dams combined divert 210,000 cfs, split between
the two countries. Designed for a hundred-year flood, this leaves
only 20,000 cfs in the main channel through the Brownsville-
Matamoros area and reduces flooding in these cities.

Given today's low flow, it is hard to imagine that steamboats
traveled the Lower Rio Grande as early as 1829. Starting in 1757,
flatboats were used for trade as far upstream as Laredo—a distance of over 350 river miles. As a result of varying river levels
and economic ups and downs, steamboats and keel boats were
used intermittently through the 1800s. In 1850 a keel boat made
it beyond the Devils River, over 500 river miles from the mouth.

The main port on the Rio Grande for cotton transport from
1850 to 1900 was the town of Roma in southwestern Starr County. Cotton farming was so prevalent in the area that by 1874 irrigation lowered the river at Roma by almost 3 feet. The first
international bridge over the Rio Grande was built in Roma in
1927. Today Roma is one of the best examples of a Spanish co-

lonial town on the Lower Rio Grande. About 30 miles downstream from Roma at Los Ebanos a hand-pulled ferry still carries cars and people across the Rio Grande. The rich history of the Rio Grande and its ties to the Spanish and Mexican heritage of Texas led to the designation of the Texas portion of the Rio Grande as an American Heritage River. Fourteen U.S. rivers have received this official designation, which helps communities to implement projects related to cultural and historic preservation, environmental protection, and economic revitalization.

Despite all the manipulations of the riverbed, diversions of water, and pollution from agriculture, cities, and industry, the Lower Rio Grande remains one of the prime birding areas in the United States as well as habitat for 1,100 types of plants, 300 butterfly species, and 216 vertebrates other than birds. The Central and Mississippi flyways pass through the area below Falcon Dam, creating an area where 484 bird species migrate or reside. As a result, the Valley is one of the world's foremost destinations for birdwatchers.

The last 275 miles of the Rio Grande below Falcon Dam were considered such a significant habitat threatened by the effects of growth in the area that Congress created the Lower Rio Grande Valley National Wildlife Refuge in 1980. This project, which consists of more than one hundred tracts and will eventually encompass 132,500 acres, is considered one of the most diverse national wildlife refuges and includes the largest remaining stand of subtropical thorn forest in the United States. Water will increasingly be an issue for these woodlands in the future. Endangered animals such as two rare felines, the ocelots and jaguarundi, are native to this area.

Brownsville is the last city on the Rio Grande before it reaches the Gulf of Mexico. It was settled in 1781, later than many towns in the area, and was slow to grow. When General Zachary Taylor established a fort there during the Mexican War in 1846, Brownsville was still a town of less than one thousand inhabitants. During the Civil War, the Confederates used Brownsville as a shipping point for cotton because of the Union blockade of other ports. Finally, in 1865, the Union Army attacked the Con-

Ruins of a Spanish Colonial mission now partially under the Falcon Reservoir. Photo courtesy of Texas Parks and Wildlife Department.

federate troops at Brownsville in what was the last battle of the Civil War—a month after General Robert E. Lee surrendered. The railroad came to the area in 1904, and a railroad bridge was built to Matamoros in 1910. This spurred the growth of the citrus industry as well as Brownsville.

A ship channel and inland deepwater port were constructed in 1936, and in 1949 the Gulf Intracoastal Waterway was extended to Brownsville, making it the southernmost port in Texas. In recent years, open-bay disposal of dredge spoils from this reach of the canal into the Lower Laguna Madre has become increasingly controversial—to the extent that some people have called for closure of the waterway. Disposal of material from dredging can cause increased turbidity, which is harmful to sea grasses. A sea grass meadow provides food for small animals, surfaces to cling to for small crawling organisms, hiding places for small invertebrates and fish, and ambush points for larger predators. In addition, there are occasional attempts to extend the canal through the lagunas of Mexico to Veracruz, though the environmental impact of such a plan would be substantial.

The Lower Rio Grande Valley is sandwiched between two growing countries. Although our ties and mutual concerns are strong, the challenges of managing the shared river will likely continue.

THE RIVERS OF THE COASTAL PLAINS

The rivers that meander across the alluvial coastal plains of Texas, especially the lower reaches of the Trinity, Brazos, and Colorado, played a significant role in the settlement of Texas. Although the Spanish did not establish any major settlements in the upper coastal area, they were the first to discover the openings to the sea in the 1500s, as shown on early maps. Robert de La Salle explored and established a colony in the 1680s on Garcitas Creek, a tributary to Lavaca Bay. When Moses Austin was given a land grant between the Lower Brazos and the Lower Colorado in the 1820s, his son Stephen established a colony that became the birthplace of the Republic of Texas. Austin's first boatload of colonists were shipwrecked and washed ashore at the mouth of the Brazos on Quintana Beach—there, Anglo Texas was born. Texas won its independence from Mexico just sixteen years later at a battle on the San Jacinto River, a tributary of Galveston Bay. Today, less than 175 years later, the city of Houston on the bay's upper reaches is the fourth largest city in the United States.

The Lower Trinity and San Jacinto Rivers

As the Trinity River begins to back up into Lake Livingston in

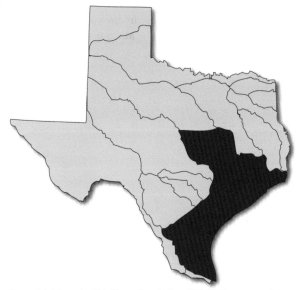

The coastal plains region. Data Source: Texas Parks and Wildlife Department and Texas Water Development Board.

the Piney Woods of East Texas, it is no longer a water provider and recipient of wastewater for the Dallas–Fort Worth Metroplex but becomes the principal source of surface water for the Houston area. Lake Livingston, with a capacity of 1,788,000 acre-feet, is the largest reservoir on the Trinity River and is owned by the city of Houston and the Trinity River Authority. Only the small Lake Anahuac lies downstream between Lake Livingston and the mouth of the river at Trinity Bay, a distance of about a hundred miles. Several ecosystems in this area downstream of Lake Livingston have been preserved as parks and wilderness areas. The 163,000-acre Sam Houston National Forest flanks the Trinity on both sides for several miles. This area also includes portions of the 84,000-acre Big Thicket National Preserve. A recent initiative known as Houston Wilderness is actively pursuing preservation of protected areas and ecotourism in the region as part of a twenty-four-county complex.

The San Jacinto River, much smaller than the Trinity, also

flows into Galveston Bay. The San Jacinto drains less than 4,000 square miles and begins just west of the Lower Trinity. There are two major reservoirs on the San Jacinto that provide water for Houston, Lake Houston, and Lake Conroe. Lower courses of the San Jacinto are used for navigation to the Port of Houston and consequently have significant water quality issues. The west bank of the San Jacinto River is famous as the location of the battle in which Texas won its independence from Mexico in 1836.

Before the railroads came to Texas in the 1860s, river navigation was common. The Trinity was used off and on into the late 1800s for river transport upstream to near Dallas. Plans for a canal linking Houston and Dallas were discussed from the late 1800s until 1973, when Dallas–Fort Worth voters refused to fund the project.

Near the mouth of the Trinity is the controversial Wallisville Saltwater Barrier Project, completed in 1999 but started in 1970. By 1973 the project to prevent saltwater from migrating upstream on the Trinity was 73 percent complete, but very sen-

The Lower Trinity River is beautiful despite Texans' dependence on it upstream. Photo courtesy of Texas Parks and Wildlife Department.

sitive ecosystems on its lower reaches were threatened. A federal judge stopped work due to inadequacies in the Environmental Impact Statement, and construction on a much-revised project resumed in 1991. Similar to the situation at Lake Sam Rayburn, freshwater had been released from Lake Livingston to push back the saltwater line when flows were low. After the construction of the Wallisville Project, these releases are no longer necessary. Today lingering questions remain concerning how the artificial stoppage of this once-fluid line of salinity will affect aquatic plants and animals in the long term.

The Trinity carries the burden of municipal supply for much of the Texas population. The city of Houston alone has permitted municipal and industrial rights for 1,258,829 acre-feet on the combined surface water sources in the area, some from interbasin transfers including the Brazos River. Over 450,000 acre-feet of water rights from the Lower Brazos have been granted to various entities in the Houston region.

The Trinity also absorbs a substantial portion of the point and nonpoint source pollution from the Houston metropolitan area. Point source pollution refers to pollution that comes from a specific source, for example, the end of a pipe. Nonpoint source pollution cannot be attributed to one specific location; examples would be runoff from urban streets and agricultural fields. Two-thirds of the petrochemical production in the United States occurs in and around Houston, and the rivers, streams, bayous, and bays are adversely affected by chemicals in the discharges from some of these facilities. In spite of the negative effects of return flows from both the Dallas–Fort Worth and Houston areas, these supplies are being evaluated as an important source of water in the future for the Trinity Basin.

Groundwater pumping, mainly from the Gulf Coast Aquifer, has been significant in the Houston area, where groundwater accounts for 34 percent of the regional supply. The overpumping of groundwater led to subsidence, an irreversible condition. The levels of the Evangeline and Chicot Aquifers dropped by as much as 400 feet by 1973, causing subsidence of up to 9 feet in parts of Harris County. The establishment of subsidence districts has enabled regulation of groundwater, and the problem has been greatly mitigated. These coastal subsidence districts were among the first groundwater regulatory entities in Texas, the only state in the nation where most groundwater pumping is still virtually unregulated.

The Lower Brazos River

The Lower Brazos River near the coast may be termed an oxbow lake. Similar to the *resacas* along the Lower Rio Grande, oxbow lakes are crescent-shaped ponds formed when the meandering river- or streambed is cut off from the main channel. The Brazos River floodplain, as it meanders near Houston, is 8 to 10 miles wide and subject to oxbow formation. A study at Texas A&M University found over thirty oxbow lakes in the middle-lower reaches of the Brazos. With no major reservoirs south of the Waco area, much of the Lower Brazos is left to meander. Brazos Bend State Park lies about 30 miles southwest of Houston and is a showplace for oxbow lakes, alligators, and premium bird-watching. *National Geographic* named it one of the nation's ten outstanding state parks.

The Lower Brazos is best known in Texas as the site of the first Anglo colony, established by Stephen F. Austin in 1821. Austin has been called the father of Texas and served as the state's first president after its independence from Mexico. Farther up the river, at the place known as Washington on the Brazos, the Texas Declaration of Independence was signed. State historic sites located here and at San Felipe celebrate the early importance of this area of the Lower Brazos in Texas history. The 250-mile stretch of the Brazos from the Gulf to Washington on the Brazos was navigable, and cotton was the main cargo until the arrival of the railroad.

The Brazos is one of the few rivers in Texas that does not empty into an estuary or bay. Instead it simply comes to a mouth in Freeport and enters the Gulf. Like many Texas rivers, it has experienced considerable manipulation at its mouth, which was relocated seventy-five years ago. One cannot talk about the mouth of the Brazos River without mentioning the Dow Chemi-

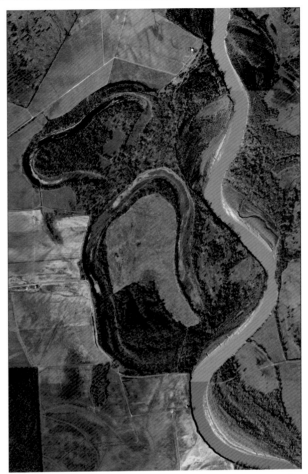

Oxbow lakes on the Lower Brazos River were formed when both ends of a natural bend were cut off from the main stem of the river. Photo courtesy of the Texas Natural Resources Information System.

cal Company. The corporation's giant chemical complex was established there in 1940 and hurriedly put on line to provide magnesium from seawater to reduce the weight of planes used in World War II. Besides millions of gallons of seawater, Dow has a permit for 280,000 acre-feet of freshwater from the Brazos, more than most cities in Texas use. In fact, in recent years water has become a profitable commodity as Dow has marketed some of its water to the growing communities of the Brazosport area.

The Lower Colorado River

After the Colorado River leaves the Highland Lakes and Austin, it becomes a wide river slowly moving through the Blackland Prairies, southern post oak savannah, and coastal plains. The LCRA controls the river's flows through management of the chain of Highland Lakes. No major cities lie downstream, even on the large estuarine complex known as Matagorda Bay. Below Lake Travis, just upstream of Lake Austin, there are no more major tributaries of the Colorado. Ninety percent of the contributing drainage occurs upstream of the Lake Travis dam. No other major reservoirs are on the Colorado downstream of Austin.

The rice farming region near the coast requires significant amounts of water at critical times that rival what some cities use. In some counties as much as 3.88 acre-feet of water per acre of land are used for rice irrigation and weed control. In 1996 Matagorda County, at the river's mouth, used 30 percent of all the water used in the fourteen-county planning region that includes Austin. Three out of the four top water-using counties in the region are heavy rice farming areas. Flood irrigation for rice farming provides an interesting side benefit to the environment as habitat for migratory birds. It is estimated that the rice prairies of the Texas coastal region provide habitat for more than two million wintering waterfowl, and the area is one of the most renowned places for waterfowl hunting in the world.

The Native Americans first encountered by European explorers along the mouth of the Colorado, Brazos, and Trinity Rivers were members of the Karankawa, a group of tribes that shared a common language. The unique characteristics and habits of the Karankawas were striking to the new explorers. The Karankawa men were tall and muscular, wore little clothing, had pierced nipples and lips, and often smeared themselves with dirt and alligator grease to ward off mosquitoes. Both men and women painted and tattooed their bodies, and women wore skirts made of Spanish moss. The tribes were nomadic and used boats made from hollowed-out trees. They followed game from the coast up the inland rivers, and fish, oysters, and turtles were their preferred food. As if their appearance was not startling enough, Karankawas were also known to have ceremonies where they would consume the flesh of their enemies. Most of the Karankawas were wiped out from contact with early Spanish and French explorers leading to epidemics and warfare with the pirate Jean Lafitte and settlers along the coast.

The name "Colorado" (meaning "red" or "colorful" in Spanish) is thought to be a misnomer that was probably meant for the neighboring Brazos River. A Spanish mapmaker most likely switched the two names. Once Stephen F. Austin established the first Anglo settlement on the Brazos, other settlers came to Texas and moved up the river to the nearby Colorado, many settling around what is now Columbus and upstream as far as Bastrop.

Trade on the Colorado was carried on from Columbus to the coast, often interrupted by floods and logjams. With the river dropping only 450 feet in 290 miles, it was a target for navigation in spite of its impediments. In 1846 a steamship was built upstream of a major logjam that blocked the mouth. This proved successful, and goods were soon delivered to Austin and another steamship was built. In 1853 the Corps of Engineers dug a channel around the logjam, which had grown to 7 miles in length. After the Civil War the new channel became blocked, and the jam grew to 40 miles long and 20 feet high. In 1928–1929 this jam was cleared and allowed to wash downstream into Matagorda Bay, where it blocked the bay navigation. In 1934 a passage was cut through the logjam. As a result, the Colorado River no longer discharged into the bay but passed through the logjam and across the barrier island to the sea. This deprived the bay of freshwater, causing long-term damage to the ecosystem. Today the river has been remodified and now discharges directly into Matagorda Bay.

Other modifications near the river's mouth include a set of locks where the Intracoastal Waterway crosses the Colorado River. During high water, these locks—the only true waterway

Parker's Cut was created in an attempt to deliver freshwater from the Lower Colorado River to Matagorda Bay. It is now cut off from the river. Photo courtesy of Texas Parks and Wildlife Department.

lock system in Texas—facilitate the travel of barges on the canal. Floodgates were originally built at this location in 1944 to reduce eroding of the waterway and the amount of dredging required. In 1952 the floodgates were converted to locks in order to make it easier to navigate during high water on the Colorado River. Near Bay City, just several miles from the river's mouth, an inflatable dam was built in the 1960s for rice irrigation pumping. That dam has been replaced with a permanent weir dam.

Southwest of the Colorado, the Lavaca River drains only 2,309 square miles before it too enters the Matagorda Bay system. Although the Lavaca basin is a small watershed, it forms the basis for an entire planning region, the smallest of the sixteen in Texas. This three-county area lies next to the counties at the mouth of the Colorado and shares many of the same characteristics, mainly substantial rice farming. However, unlike irrigation on the Lower Colorado, groundwater pumping provides 90 percent of the water supply for the fields along the Lavaca. Approximately 96 percent of total water use here is for irrigation as there are no major cities in the watershed or on Matagorda Bay.

There is one reservoir on the Lavaca River, Lake Texana. From this lake, which has a capacity of 79,000 acre-feet, 42,000 acre-feet are contracted for municipal use to Corpus Christi, which is two basins away at the mouth of the water-starved Nueces River. A 102-mile, 64-inch pipeline delivers this interbasin transfer. By virtue of an agreement among the Sierra Club, the state of Texas, and the Lavaca Navidad River Authority, 4,500 acre-feet of Lake Texana are reserved for environmental flows to Matagorda Bay.

As it enters the bay through its rerouted path, the Colorado River passes through jetties made of granite boulders shipped down from the ancient granite uplift in the Hill Country near the Highland Lakes. The imprint of development has been heavy on this river system, and this is sure to persist as interests throughout the basin and beyond continue to increase demands on its flow.

In 2004 LCRA began a six-year feasibility study of plans to help meet long-term future water needs by capturing stored or unused Colorado River flows in an off-channel storage facility and conveying that water to San Antonio without harming the Colorado River or Matagorda Bay resources. The project is looking at reducing agricultural irrigation demands for water with such conservation measures as growing more water-efficient rice. The project is also studying the use of groundwater for agricultural needs when surface water is not available.

An independent scientific review panel is reviewing the studies and their findings. As a result, LCRA included global climate change, impacts to waterfowl, and other considerations as factors in their studies. LCRA has also held meetings to inform the public of their progress and to gather feedback from stakeholders. For example, local residents and community leaders told LCRA that finding a willing seller was an important consideration. In 2007 LCRA identified landowners willing to sell a 4,200-acre portion of their ranches in Wharton County for the site of the off-channel storage facility.

Based on the study results, the LCRA and the San Antonio Water System (SAWS) boards will decide whether to implement the project in 2010. The boards have agreed that the plan will only be implemented if the project meets legislative requirements and is technically, environmentally, and financially feasible.

The Lower Guadalupe and San Antonio Rivers

THE LOWER GUADALUPE RIVER The Guadalupe River changes from a Hill Country stream of rapids and pools to a slower, more meandering river as it begins its journey through the coastal plains to San Antonio Bay. Victoria, on the Lower Guadalupe, is the largest city on the Guadalupe River with 84,000 people. The next largest towns are New Braunfels and San Marcos, each with a population of about 40,000. Besides the spring-fed San Marcos and Comal Rivers, there are no major tributaries for approximately 250 miles until the San Antonio meets the Guadalupe about 10 miles inland from the coast. The Comal and San Marcos Springs can provide up to 80 percent of the flow to the lower areas of the Guadalupe River in a serious drought. Ranching and agriculture combined with oil and gas production make up much of the economy of the Lower Guadalupe region.

In addition to the hydropower generation complex at Canyon Dam, five small hydropower facilities take advantage of the continuing drop of the river between New Braunfels and Cuero. These are mainly owned by the GBRA now, and their already limited hydro-potential is further reduced as more water is transferred out of Canyon Lake to surrounding communities, thereby reducing average flows in the Guadalupe.

The Victoria area, although a center for ranching, oil, and gas, has also become home to several major industries, due in part to the availability of water from the Guadalupe River. DuPont, Dow Chemical, B. P. Chemicals, and Formosa Plastics have plants there. Two of these companies, Dow and DuPont, own large water rights downstream of Victoria. These rights and the Guadalupe Blanco River Authority rights on the Lower Guadalupe comprise almost 80 percent of all the water rights on the Guadalupe River. The upper and middle portions of the Guadalupe have relied on the Edwards Aquifer, and thus no substantial surface water projects, except for Canyon Lake, have been developed there. One reason for the lack of surface water devel-

The Port of Victoria depends on the Lower Guadalupe for freshwater. Data Source: Port of Victoria.

opment in the upper reaches is that, in contrast to the Victoria area industrial complex, no major industry exists in the upper and middle Guadalupe.

Some attempts at river navigation on the Guadalupe were made in the early 1800s, but snags and low flows often interrupted traffic. The Guadalupe, a more western river in Texas, has even more low flow periods than its eastern neighbors. In the early 1900s Victoria businessmen were instrumental in efforts to create the Gulf Intracoastal Waterway, which by 1949 eventually stretched all the way from Sabine Lake to Brownsville. A side channel 35 miles long linking Victoria and the Intracoastal Canal was built in the mid-1960s. The barge canal basically parallels the Guadalupe and enters San Antonio Bay at a different point than the main stem of the Guadalupe. There is concern that the amount of freshwater moving down the canal instead of the river channel has affected the ecology of San Antonio Bay, including the sedimentary deposits in the delta. As in other Texas rivers, the mouth of the Guadalupe has been altered with an adjustable saltwater barrier. This inflatable dam can be raised or lowered depending on the movement of the saltwater line, thus maintaining a dependable supply of freshwater for the Victoria area upstream.

THE LOWER SAN ANTONIO RIVER About 10 miles before entering San Antonio Bay, the Guadalupe is joined by the San Antonio River. The latter, though smaller, contains a much larger population in its basin, concentrated in metropolitan San Antonio, which has 1.5 million people. Consequently, and because they are part of the same system, there is considerable pressure from San Antonio River interests to obtain water from the Guadalupe. Inconveniently and also due to politics, the state legally classifies these two rivers as totally separate basins. Therefore, any water moved between the two is subject to the rules of interbasin transfer. Under these rules, any water moved becomes junior to all other rights on the river of origin. That means that in the event of a drought and mandatory restrictions of water, the junior water right, or most recently permitted, would be the

In 2004 over 5,700 barges transported goods through the Port of Victoria. Photo from iStock.com.

first to have to stop pumping. The newest right would be the interbasin transfer right. Because of the junior priority of these interbasin transfer rights, they are not very desirable or valuable to entities such as cities that need dependable supplies. Efforts have been made in the Texas Legislature by interests in the city of San Antonio and associated groups to reclassify the San Antonio River and the Guadalupe River as one basin. To date, that hydro-political effort has been unsuccessful.

The San Antonio River downstream of the city of San Antonio is rural, fairly narrow, winding, and small and lined with dense brush. Farming and ranching are the major activities, and no major towns or cities are on the river below San Antonio. In recent years, some segments of the Lower San Antonio have been classified as failing to meet the standards for contact recreation and have been placed on the Texas 2002 Clean Water Act Section 303(d) list. Fecal coliform levels have exceeded the maximum allowed for swimming areas. Ongoing studies are attempting to pinpoint the sources of the contamination but are thus far inconclusive.

Downstream rural areas blame poor wastewater treatment plants in San Antonio. However, tests show that the areas near San Antonio generally meet contact recreation standards, and many of the older wastewater plants were improved in the late 1980s. The fecal counts increase with rainfall events, according to the San Antonio River Authority. The critical areas are miles downstream of the city after the confluences of several creeks, leading to the assumption that the high fecal coliform readings are due to nonpoint sources, including urban runoff and concentrated animal feeding operations. Laboratory tests show that the origin of most of the fecal coliform in the lower Guadalupe is animals rather than humans.

THE GUADALUPE ESTUARY Although San Antonio Bay has only one community, Seadrift, with a population of 1,000, the stretch near the river's mouth where the Guadalupe and San Antonio Rivers converge is currently under increasing pressure. Plans were proposed for a 120-mile pipeline from the saltwater barrier back to San Antonio, carrying an estimated 94,500 acre-feet of river water. To supplement the Guadalupe in a record drought causing little or no flow from Victoria to the bay, the GBRA proposed using groundwater from the lower basin to meet San Antonio's needs. The strategy produced significant opposition in the region and beyond. Ranchers and farmers from Victoria to the coast feared their groundwater would be depleted. Victoria area residents were worried that their emergency supply groundwater would not be available in a drought and would be sent upstream to San Antonio. Anglers, hunters, recreational users, shrimpers, and environmentalists were concerned about the reduced river flow and freshwater inflow to San Antonio Bay. The bay adjoins the winter home of the endangered whooping crane, and studies are under way to determine the freshwater needs related to that species.

Originally the San Antonio Water System was a major proponent and funder for this project. However, in 2005 SAWS decided to withdraw from it. The GBRA reconfigured the project without the controversial groundwater component. Still, many concerns about this surface water project remain, including impact on San Antonio Bay fish and wildlife and commercial and recreational fishing industries. In spite of opposition, this modified plan is included in the 2007 State Water Plan. A large ranching interest near Victoria filed a lawsuit at the end of 2006 against the Water Development Board claiming that the project should not be included in the Region L plan.

THE GUADALUPE—FLOWING BUT THREATENED With only one major dam and no major cities on its banks, the Guadalupe is a natural stream in many of its reaches. In the Texas Water Safari, paddlers race 268 miles from the Texas State University campus in San Marcos down the San Marcos River to the Guadalupe River and then to Seadrift on the bay, taking advantage of the relative natural state of the river. Yet, as the Guadalupe Basin feels pressure from the growing San Antonio and Central Texas area, this free-flowing river may not survive much longer in its seminatural state.

The Texas Water Safari, a 262-mile canoe race, is considered "the world's toughest boat race." Photo by Paula Goynes.

The Lower Nueces River

NUECES TRIBUTARIES The lower portions of the Nueces and its tributaries, the Frio, Sabinal, and Atascosa, flow through dry South Texas brush country vegetated with mainly mesquite and cactus. Oil and gas operations are carried out in many of the counties along this lower reach. About 100 miles from the coast, the Frio River is impounded at Choke Canyon Reservoir just before its confluence with the Nueces. Just below the reservoir in the town of Three Rivers, the Atascosa and Nueces Rivers join the Frio and together flow 63 miles before reaching Lake Corpus Christi. About 30 miles below Lake Corpus Christi is Calallen Dam, a small dam built to stop saltwater intrusion and to provide an intake for municipal water.

THE NUECES RIVER AND CORPUS CHRISTI Water used from the Nueces and Frio Rivers is mostly for municipal and industrial purposes in the Corpus Christi metro area, which has a population of 382,000. The area is known as the Coastal Bend Region and consists of twelve counties extending from Corpus Christi to near Brownsville and has a total population of 550,000. There are

267 water rights in the Nueces Basin, and two of those rights for the Corpus Christi area make up 83 percent of the total. Choke Canyon Reservoir, Lake Corpus Christi, and Calallen Reservoir are operated as a system to provide Corpus Christi with water. Additional water is available from interbasin transfers, including 46,000 acre-feet from Lake Texana in the Lavaca River Basin and 35,000 acre-feet from the Colorado River. Water from Lake Texana is delivered through a 102-mile pipeline from the Lavaca River Basin. The Colorado River rights have not yet been used.

The Corpus Christi area has the third largest refinery and petrochemical complex in the United States. To their credit, the plants here use, on the average, 60 percent less water to refine a barrel of crude oil than refineries in the Houston-Beaumont area. Some of these refineries reuse water up to fifty times before discharging.

One of the challenges in providing water in this dry, hot area of the state is evaporation loss. As water is stored in open lakes and moves down the open Nueces River channel to Corpus Christi, a significant portion is lost to evaporation, transpiration, and seepage along river channels. Estimates of peak losses are 29 percent from Choke Canyon to Lake Corpus Christi plus an average of 7 percent more from Lake Corpus Christi to Calallen Dam. This is a significant loss of water from a dry region. More disturbing for the area, evaporative losses in the future are expected to increase by 30 percent in normal conditions and 85 percent in drought conditions as a result of climate changes. However, due to limited growth projections and added availability through the interbasin transfers, the region projects a shortfall of only 1,000 acre-feet by 2050.

Groundwater from the Gulf Coast Aquifer will make up a portion of the overall need. The Carrizo-Wilcox Aquifer spans Texas from Arkansas to the Rio Grande and passes underneath three counties in the Corpus Christi area. It is interesting to note the variance in temperatures in this area: in some places, the temperature of the water is 140 degrees F, whereas about 200 miles away at College Station, the water from the Carrizo-Wilcox is 118 degrees F and is used for municipal purposes after cooling. Groundwater in the area of Calallen Reservoir is thought to contribute to the excessive chlorides found in the lake water. Excessive minerals such as chlorides cause buildup of mineral deposits in industrial cooling facilities and subsequently require additional water for cleaning and dilution. Blending water from the Colorado and Lavaca Rivers 100 miles to the northeast is one proposed solution to the industrial water problem.

The major water controversy in the Corpus Christi area is freshwater inflow from the Nueces River. A specified amount has to be passed through Lake Corpus Christi to the Nueces Bay unless the reservoir drops below a certain level—at which point municipal needs trump environmental needs. There is much discussion about the inadequacy of the target flows that have to be released for the estuary, as well as the portions of the bay into which the Nueces River flows. A joint project of the Nature Conservancy, the Texas Parks and Wildlife Department, and the city of Corpus Christi has improved the point-of-discharge problem by redirecting some flows to critical habitat areas rather than directly into the bay.

The port at Corpus Christi is the fifth largest in the United States in terms of tons of goods shipped. This is even more impressive given that the Corpus Christi Bay was one of the last areas of Texas to be settled, remaining uninhabited until 1839, after Texas gained its independence. A German nobleman tried to establish a colony on Nueces Bay in the 1830s, but a blockade by the French because of its relations with Mexico kept the colonists from landing. Lack of a deepwater port stymied growth until 1926, when a ship channel and port were completed.

The Nueces River is essentially Corpus Christi's river. While allowed to run wild and free in its many Hill Country tributaries, it is eventually harnessed in the three reservoirs and fully utilized to maintain this growing South Texas industrial, municipal, and recreational center.

TEXAS-SIZED RANCHES Between the Nueces Basin along the coast and the Rio Grande Basin lies a unique area of Texas, historically, hydrologically, and biologically. This 100-mile area of

The King Ranch is famous for its Santa Gertrudis cattle, its age, and its size. Photo courtesy of Texas Parks and Wildlife Department.

coast consists mainly of two counties and a few ranches that are among the oldest in Texas. One of Texas' most famous ranches is the King Ranch, founded in 1852. The King Ranch is associated with the development of the Santa Gertrudis breed of cattle. Covering 825,000 acres and several counties, the ranch has holdings worldwide. Kenedy County, whose population is 408, sits in the middle of this isolated portion of the state. There have never been more than twenty-five ranches in this county of almost 1,400 square miles.

No year-round sources of freshwater exist in the area, and this lack of water is in part responsible for the high salinity in Laguna Madre, the only lagoon in the United States saltier than the ocean and one of only five hypersaline lagoons in the world. It is also home to hundreds of thousands of migrating birds, including one-third of North America's peregrine falcons. The Laguna has the most extensive sea grass beds in Texas, and this contributes to a healthy sport fishery that is world famous. The most significant human alterations in the area are a pass through the 100-mile barrier island near Baffin Bay and the Intracoastal Waterway from Corpus Christi to Brownsville. With no real freshwater inflow from rivers or streams, the Laguna Madre carves out its unique ecological niche in one of the remotest parts of Texas.

Peach Point Wildlife Management Area is located in Brazoria County near the Texas coast. Photo courtesy of Texas Parks and Wildlife Department.

4. THE GULF SHORES OF TEXAS

FRESHWATER, SALINITY, AND BARRIER ISLANDS

Freshwater comprises merely 2.5 percent of the world's water supply, and of that only 0.3 percent is available in rivers and streams. Approximately 30.1 percent of freshwater is held in groundwater, and 68.7 percent is held in ice caps and glaciers. The oceans hold 97.5 percent of the Earth's water. Freshwater enters the oceans in various forms—as river flow, as runoff from rain falling on the shore, as seeps from underground springs, as ice melt, and as direct rainfall. At the oceans' edges are complex, moving, ever-changing blends of fresh- and salt water that make the delineation of the two almost impossible.

The mixing of freshwater with the sea varies greatly according to geology, hydrology, climate, tidal movement, and human alterations of the rivers and coastlines. Where rivers flow directly into the sea without natural obstructions such as barrier islands, marshes, or wetlands, mixing of fresh- and salt water is limited. However, in complex systems like the Texas coast, with

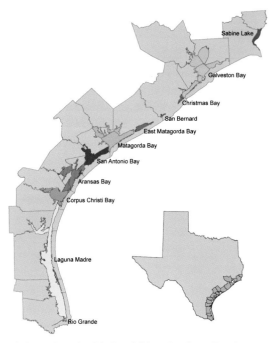

The major bays and estuaries of the Texas Gulf Coast. Data Source: Texas Commission on Environmental Quality.

its submerged vegetation, almost continuous barrier islands, and only a dozen or so narrow connections to the sea, there is considerable retention time and space to form zones of varying salinity. The key geographic features of these complex systems are bays and estuaries. A bay is simply an indentation of a shoreline larger than a cove. An estuary is a partially enclosed body of water where freshwater inflows from rivers and streams meet the ocean and mix with seawater. The terms *estuary* and *bay* are often used in tandem, as a large complicated estuary system like the Colorado River estuary can have several bays within it. The majority of the rivers in Texas flow into bays. Only the Brazos, San Bernard, and Rio Grande flow directly to the Gulf of Mexico without a bay or estuarine system.

The zones of varying salinity in the estuaries and their accompanying habitat, including sea grasses and marsh plants, form some of the most productive ecosystems in the natural world. An acre of salt marsh in an estuary can produce six times more organic matter than an average acre of wheat-producing farmland and twenty times more than the ocean. An amazing 95 percent of the Gulf of Mexico's important recreational and commercial fish and shellfish spend at least part of their life cycles in

Saltwater marshes, like this one at Sabine Pass, are among the most productive ecosystems in the world. Photo courtesy of Texas Parks and Wildlife Department.

the estuaries. Oysters, which are an important commercial fishery in Texas, can thrive only in estuaries as they are attached and immobile and depend on a specific mix of fresh- and salt water. Galveston Bay is the nation's most productive oyster habitat. Most other species that live in the estuary are spawned in the Gulf, and their tiny planktonic offspring drift in on the tides to find refuge in estuarine nursery areas, where salinity is lower and there is protection and nutrition from marsh or sea grass vegetation. In this environment, species such as redfish and speckled trout can mature in a zone where their predators do not normally prowl and where there are abundant microscopic food sources on which they depend in early stages of life.

Texas has seven barrier islands comprising over 300 of its 450-mile coastline. Only Florida and Alaska have more miles of islands along their coasts. The longest island of our barrier islands is South Padre Island, stretching 113 miles from Corpus Christi to Brownsville. Two types of estuaries form behind these barrier islands—drowned river mouths and bar-formed lagoons. Drowned river mouths, submerged by the ancient rising sea, are found at Sabine Lake, Trinity-Galveston, Lavaca-Matagorda, and Nueces-Corpus Christi Bays. Bar-formed lagoons are found at east and west Galveston Bays, east Matagorda Bay, the Aransas and Redfish Bays, and the Laguna Madre.

The structure of the estuary and its barrier islands and the location and size of the openings to the ocean have a major effect on the range of salinity zones, but the amount of freshwater coming down rivers and streams is the main determinant. For example, the Laguna Madre Estuary has no perennial rivers or streams flowing into it. Thus its salinity is very high. In fact, the Laguna is one of only five hypersaline lagoons in the world. This high salt condition is increased by the shallowness of the estuary and the high ambient temperatures that increase evaporation. At the opposite end, both geographically and in terms of salinity, is the Sabine-Neches Estuary on the extreme northeastern coast of Texas, which receives more freshwater inflow than any other: more than fifty times its average volume annually.

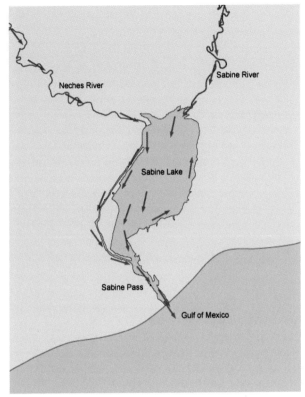

The Sabine-Neches Estuary has the lowest salinity in Texas and is a tidal lake. Red arrows indicate freshwater flow during low tide. Data Source: Texas Commission on Environmental Quality.

Galveston Bay and San Antonio Bay receive four to seven times their volume annually, while the Nueces and Mission-Aransas Bays receive only 60 to 70 percent of their volume.

The tides of the Gulf of Mexico generally do not vary more than about 3 feet annually, but they do affect the salinity of the bays. Tides alternately push saltier water upriver and then pull the wedge, or line of salt water, back to the Gulf through narrow passes. Texas estuaries are generally shallow and protected by barrier islands, so wind has a greater influence on the circulation of water in the bays than the tides.

Human alterations to the shoreline and bottom of the bays can affect salinity as well. The Houston Ship Channel in Galveston Bay, for example, is dredged up to 45 feet deep and 530 feet wide from the open Gulf through a bay that is normally less than 12 feet deep. The channel creates a funnel, bringing saltier water into the bay at rates and volumes greater than those that occur naturally. Seawalls and breakwaters placed at the edges of estuaries have replaced wetland marshes and the natural mixing of fresh- and salt water at shallow depths in beds of ecologically important sea grasses. Man-made saltwater barriers stop migration of the salt wedge up river channels, reducing the habitat for estuarine plants and animals. The Intracoastal Waterway also allows salt water to move in pathways that did not exist naturally, causing more salinity in some areas or, in the case of Laguna Madre, causing more freshwater to enter the lagoon.

The human actions that have had the greatest effect on the salinity balance of the estuaries are water diversion and storage in rivers and reservoirs. Although many of the rivers in Texas are overappropriated, these rights have generally not been fully utilized and freshwater inflows have continued. In the future as greater pressure is placed on these historic rights, especially in dry years, there will be little or no freshwater inflow into the bays and estuaries, and they may become too saline for many species to survive. Despite the 2007 passage of Senate Bill 3, which addresses protection of freshwater inflows to the bays, there are still no provisions related to existing water rights in overappropriated basins.

Water that is released from reservoirs at times when natural flows do not occur can also alter the salinity of an estuary at key periods during the year. Many of the estuarine plants and animals have adapted to seasonal salinity fluctuations, and their survival is dependent on those cycles. Detailed studies by the Texas Parks and Wildlife Department and the Texas Water

Development Board are ongoing to determine the minimum necessary amounts and timing of freshwater that each estuary needs. Thanks to years of research, Texas knows more about the needs of its bays and estuaries than most other places in the world. However, the bay and estuary needs studies have drawn some criticism because of questions on methodology and on applicability to low flow situations. To address this issue, the state legislature passed Senate Bill 3, which laid out a new program to establish a specific flow regime for all the major estuaries and river basins in the state. A scientific team will be selected for each major bay and river basin to study and make recommendations on environmental flow requirements. These recommendations, along with input from stakeholders and state agencies, will be reviewed by the TCEQ for approval. The first priority group, consisting of the Sabine Lake and Galveston Bay systems, has a September 2009 deadline for proposed flow regimes.

THE GULF OF MEXICO AND OCEAN CURRENTS

The Gulf of Mexico was the gateway that first brought explorers and settlers to Texas. It was not just coincidence that the Spanish ended up in the Caribbean, then in Mexico, and finally in Texas. Gulf currents move in a circular pattern of water circulation that brought ships up the passage between Cuba and the Yucatán to Veracruz and then along the Texas coast to Florida and into the Gulf Stream, facilitating a round-trip voyage from Europe—for those who survived. This same circular current helps to wash ashore more than 500 tons of trash each year on the beaches of the midcoast of Texas.

Early Spanish explorers rarely traveled beyond the barrier islands. One of the reasons was the fierce tribe of coast-dwelling Karankawa Indians that inhabited the islands from Galveston to Corpus Christi. Only after attempted settlement by the Frenchman La Salle in 1685, more than one hundred years after the first European, Cabeza de Vaca, washed up near Galveston, did the Spanish begin to explore inland to defend their possessions from the French.

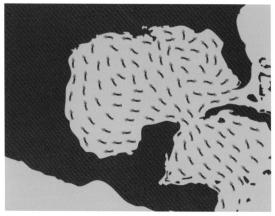

The Loop Current in the Gulf of Mexico. Data Source: Cooperative Institute for Marine and Atmospheric Studies.

THE GULF INTRACOASTAL WATERWAY

More than any other man-made project, the Intracoastal Waterway has been responsible for the profoundest changes in Texas bays and estuaries. The Texas reach of the canal extends 426 miles from Sabine Pass to the Brownsville Ship Channel at Port Isabel. The Sabine–Corpus Christi segments were completed in 1941, and the Brownsville segment was completed in 1949. While the concept of an intracoastal canal dates back to 1808, the length of time it took to complete it is blamed in part on the railroads, which lowered their rates and sometimes refused transfers of freight from waterways.

The entire inland passage goes from the Chesapeake Bay to Brownsville and passes through man-made canals and coastal lakes and behind barrier islands, providing a protected barge route for the movement of cargo. There were nearly one hundred thousand one-way trips in Texas in 2002, most consisting of multiple barges limited to five barges per tow. These barges moved 63 million tons of bulk materials, including petroleum products, chemicals, iron and steel, building materials, and sulfur, collectively worth $25 billion.

Transport by barge is considered the most economical and

safest way to move goods. Comparative fuel consumption ratings of gallons of fuel to move one ton of cargo in 2002 were as follows:

- Truck—1 gallon of fuel moves 1 ton 60 miles
- Rail—1 gallon moves 1 ton 202 miles
- Barge—1 gallon moves 1 ton 514 miles

In addition, spills are a low risk in barge traffic accidents. There were only two barge spills in 2002 in the Texas reach, as compared to 124 rail spills and 1,027 spills from truck accidents. On the other hand, a barge can carry much more than a railcar or truck, so a sizable spill could be disastrous. One barge = 15 railcars = 60 trucks. The potential for a large spill is of particular concern where the Intracoastal Waterway passes through sensitive habitats, including the winter home of the endangered whooping crane at Aransas National Wildlife Refuge.

The average dimensions of the waterway are 12 feet deep and 125 feet wide. In bays and lagoons, including the Laguna Madre, which average only a few feet in depth, the waterway can have a significant impact on currents, water temperature, wetland

The Gulf Intracoastal Waterway extends 426 miles along the Texas Coast. Photo courtesy of Texas Parks and Wildlife Department.

areas, and other habitat characteristics. Constant dredging is required to keep these channels open. In 2003 almost 8 million cubic yards of material were dredged for canal maintenance. Of this amount, 4.6 million cubic yards were placed in confined areas, 2.1 million were simply dispersed into the bays, and one million cubic yards were put to beneficial use (i.e., the material was used as a resource instead of treating it as waste). Some beneficial uses are wetlands creation, waterbird nesting and roosting islands, offshore reefs, and construction aggregate.

In 1994 environmental groups and sportfishing enthusiasts challenged open-bay disposal of dredge spoils in a lawsuit against the Corps of Engineers, asking for an environmental impact statement on disposal procedures in the Lower Laguna Madre. A report was issued in 2003 calling for more beneficial usage of the dredged material to reduce open-bay disposal. These groups noted that the usage of the Intracoastal Waterway from Corpus Christi to Brownsville is very low compared to other portions of the Texas waterway.

According to Environmental Defense, only 2.5 percent of all Texas Intracoastal Waterway traffic is on this section, and of this traffic, 80 percent is headed south, leaving returning barge traffic largely empty. Periodically, there have been efforts planned to extend the canal into Mexico and, conversely, efforts to replace the lower reach with pipelines to Brownsville. To date, neither has gained much support or momentum, and, unfortunately, open-bay disposal continues principally due to opposition from private landowners along the canal who believe that receiving dredge spoil would be harmful to their property.

A TOUR OF TEXAS BAYS AND ESTUARIES

Texas has one of the most unusual coastlines in North America. Our tour of the major estuaries of the Texas coast beginning with the Sabine-Neches Estuary in the north and ending with the Laguna Madre in the south will look at the history, geology, economics, demographics, and flora and fauna and the challenges facing these complex ecosystems.

The Sabine-Neches Estuary

The Sabine and Neches Rivers enter Sabine Lake, which is an inland estuary connected to the Gulf by Sabine Pass. Through this pass, over 75 million tons of cargo are moved to and from the ports of Beaumont, Port Arthur, and Orange lying several

Sabine Lake Estuary at the confluence of the Sabine and the Neches Rivers. Photo courtesy of the Texas Natural Resources Information System.

miles upstream of the estuary. The Sabine River was heavily used for navigation in the 1800s, and Port Arthur and Orange were considered key ports for exporting Texas cotton. During the Civil War, the Union attempted a naval assault on Sabine Pass. A small contingent of Confederate soldiers with six cannons held off five Union battleships at the Battle of Sabine Pass. This was one of only two Civil War skirmishes in Texas.

The Sabine River provides about 45 percent of the freshwater for Sabine Lake, while the Neches River provides 35 percent. The remaining flow is from smaller coastal streams and runoff from adjacent lands. The Sabine and Neches Rivers, in the wettest part of the state, provide substantial inflow for the estuary, averaging fifty times the volume of the estuary per year. By comparison, the Nueces Estuary to the southwest receives an average inflow of only 60 percent of its volume per year. The enclosed estuary of Sabine Lake is unique in Texas, being the smallest of the major estuaries while receiving the most freshwater inflow. It has the character of both a freshwater lake and an estuary where occasionally redfish, flounder, and black bass are found in the same environment.

In spite of the large inflows to the Sabine Estuary, the timing of the flows has concerned scientists. Due to management of two large reservoirs upstream used principally for hydropower, Toledo Bend on the Sabine and Sam Rayburn on the Neches, average low flows are higher and peak flows into the estuary are lower. The result has been lower than historic salinities in the summer, which is believed to have adversely affected invertebrates, especially shrimp. Recently, a saltwater barrier was installed near the mouth of the Neches to protect freshwater intake structures upstream. Freshwater previously had been released from Lake Livingston to push salt water downstream. There will certainly be effects on the ecosystem from both of these actions, but full understanding of their impacts will take years. There are many other factors affecting the salinity of Sabine Lake, including the Intracoastal Waterway, ship channels from the Gulf, and widening of Sabine Pass.

Erosion is a concern to the Sabine Estuary as well as to the

State Highway 87 was destroyed due to coastal erosion and has been closed since 1990. Photo courtesy of TexasFreeway.com.

Gulf shoreline adjacent to jetties that protect the mouth of Sabine Pass itself. In the Sabine Basin, a combination of land subsidence and sea level rise has resulted in a water level increase averaging 0.25 inch per year, which has resulted in the loss of adjoining wetlands. Cessation of lateral sand movement on the coastline due to the jetties, along with decreased sediment loads from the Mississippi River, has caused the beach to recede. This has been exacerbated by recent hurricanes, including Rita and Katrina in 2005. State Highway 87 from Sabine Pass along the coast toward Galveston is being consumed by the Gulf and is now closed.

Galveston Bay System

Galveston Bay is a system of contrasts. As the recipient of more than 60 percent of the wastewater in Texas, due to its location downstream from Dallas–Fort Worth and Houston, this bay system still manages to provide one-third of Texas' commercial fishing and half of its recreational fishing revenue. It is home to the Port of Houston, the second largest port in the United States, and the region accounts for half of the nation's petrochemical manufacturing and one-third of its petroleum refining. Mean-

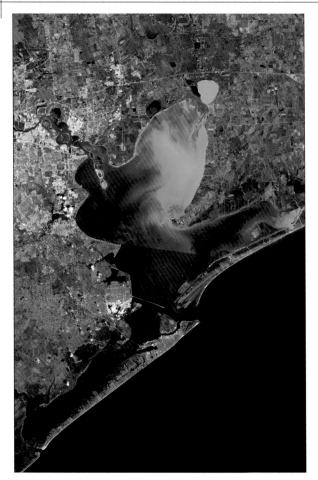

Most of Galveston Bay's freshwater comes from the Trinity and the San Jacinto Rivers.
Photo courtesy of the Texas Natural Resources Information System.

while, three-fourths of the bird species in North America spend some time in the Galveston Bay area. The bay ranks second nationally in seafood production behind the Chesapeake, and for every pound of finfish caught commercially, more than six pounds are caught by recreational fishermen, creating a combined annual revenue of $1 billion.

Galveston Island, at the mouth of the bay, is the best natural port site between Veracruz and New Orleans and was used by privateers and pirates in the early 1800s, including Jean Lafitte. Galveston grew quickly in the mid- and late 1800s, becoming the second largest city in Texas and the second largest port of entry for immigrants behind New York City. The devastating hurricane of 1900 killed between 6,000 and 12,000 people and made investors leery of a place that had been hit by eleven hurricanes in the nineteenth century. Galveston has never really recovered and may serve as an indicator of what lies ahead for New Orleans after Hurricane Katrina. Meanwhile, Houston surged ahead and built a ship channel by 1914, passing up Galveston. Today Galveston is twenty-ninth in Texas in terms of population, and Houston is the fourth largest city in the United States.

FRESHWATER INFLOW AND TIDES Freshwater inflow and water quality in the bay are of concern to scientists, environmentalists, fishermen, and ordinary citizens. The Trinity and San Jacinto Rivers are the two main freshwater contributors to Galveston Bay, with the Trinity providing 54 percent of the flow and the San Jacinto 28 percent. The Trinity provides most of the water for Dallas–Fort Worth and Houston. In spite of upstream diversions for almost 50 percent of the population of Texas, the bay system generally receives adequate flows for three reasons:

1) About 50 percent of diverted water upstream is currently returned as treated wastewater to the Trinity River. This condition is now at risk, as cities and river authorities are looking at these return flows as a new source of water.

2) Much of the appropriated upstream water has not been fully utilized. Inevitably, it will be, and at that time freshwater inflows will decrease.

3) Runoff from impervious cover, including roads, parking lots, and rooftops in the Houston area, though laden with contaminants, contributes flow in excess of predevelopment levels.

There are three tidal inlets to Galveston Bay, with 80 percent of the exchange of fresh- and salt water coming in through the pass between Galveston and Bolivar Peninsula. On the western end is San Luis Pass where most of the remaining exchange occurs and where shallow depths ensure that only recreational and commercial fishermen can access the Gulf. Minor tidal exchange occurs at Rollover Pass, cut through Bolivar Peninsula in 1955 by what was then called the Texas Game and Fish Commission, predecessor of the Texas Parks and Wildlife Department. The pass was created to improve fishing in East Galveston Bay by increasing salinity, promoting growth of submerged vegetation, and providing an additional route for spawning marine fish. The name "Rollover Pass" came from the practice of ship captains who, when Texas was under Spanish rule, would avoid customs at the Galveston Port by unloading cargo and rolling it over the narrowest point on the peninsula.

CHALLENGES FACING THE BAY AND ESTUARY

Water quality is always of concern in an estuary, especially one that adjoins the fourth largest city in the United States and has

Galveston Bay suffers from extreme pollution due to its adjacent cities and industry.
Photo courtesy of Texas Parks and Wildlife Department.

half of the nation's petrochemical manufacturing. Although it is safe to eat the fish from most parts of the state, the Texas Department of Health (TDH) has issued consumption advisories for parts of Upper Galveston Bay, the Upper Houston Ship Channel, and the San Jacinto River. Generally these advisories call for limited consumption—one 8-ounce serving per week—of catfish, blue crabs, and, in some areas, all fish. Women who are pregnant or nursing and children should not consume any of the species mentioned from these areas.

Among the chemicals found in the contaminated areas are pesticides, PCBs, and dioxins. The presence of fecal coliform bacteria in some areas of the bay has resulted in bans on oyster harvesting. Despite the periodic bans, oyster fishermen in the bay have had great success in recent years. On the brighter side, a consumption advisory for Clear Creek, a tributary of Galveston Bay, has been lifted. For ten years the creek had contained harmful levels of dichloroethane and trichlorethane, two cancer-causing chemicals. Overall, the bay is seeing a general trend of water quality improvement.

Another threat to Galveston Bay's health could come from

Located in Freeport, the Texas Operations is Dow Chemical's largest site. It manufactures 44 percent of Dow's products sold in the United States. Photo courtesy of Texas Parks and Wildlife Department.

the recently completed expansion of the Houston Ship Channel from 40 to 45 feet deep and from 400 to 530 feet wide. This is in a bay that has an average depth of about 7 feet and the ship channel itself is 51 miles long. There is concern that the wider and deeper channel to the sea will increase the salinity of the bay. Groups led by the Galveston Bay Foundation successfully negotiated a reduced scope from the original proposed dimensions of 50 feet deep and 600 feet wide. The U.S. Army Corps of Engineers agrees that 118 acres of oyster beds would be displaced by the project. They have prescribed the building of artificial oyster beds to mitigate this impact, and their studies show that oysters from these beds will mature and be edible. Another concern surrounding this project was the amount of dredged material that would be removed and where it would be deposited. Ultimately all the material was used beneficially or deposited into confined upland areas.

BAY AND WETLAND PRESERVATION AND RESTORATION As recently as the early 1990s there were about 142 miles, or 108,000 acres, of wetlands in the form of marshes that skirted the shores of Galveston Bay. Unfortunately, they are gradually being filled in by development or altered as a result of the subsidence from overpumping of groundwater from the aquifers near the coast. At least 35,000 acres of marshes have been lost. Significant wetlands restoration is being carried out by several government and nonprofit organizations. Scenic Galveston's work can be seen in the form of a 70-acre tidal marsh preserve where there was once a landfill. The Galveston Bay Foundation is working on wetlands restoration in the whole bay area, with the goal of 24,000 acres by 2010.

Galveston and Corpus Christi Bays are two of twenty-eight bays in the United States designated as units of the National Estuary Program (NEP). This federal program provides funding for planning and research to improve and maintain water quality in estuaries for fish and wildlife protection and to provide recreational opportunities. Galveston Bay is a surviving ecosystem facing a constant struggle in an area significantly affected by human activity. There is evidence that the bay's water quality is improving, despite the many threats. Its freshwater inflows are similarly threatened. Its fisheries are surviving, but at the same time its shorelines are receding and marshes decreasing. Galveston Bay NEP, established in 1998, has developed a long-range management plan for this system to address many of these resource issues. Together with other interest groups, they are monitoring the state of the bay and are pushing for protection measures in the areas of concern, as well as undertaking restoration projects. Whether they can keep pace with the multiple impacts of this fast-growing area remains to be seen.

The Lavaca-Colorado Estuary

The Lavaca-Colorado Estuary is the second largest estuary in Texas and is composed of several bays. Matagorda, East Matagorda, and Lavaca are the main bays, and the Colorado River, which now flows into West Matagorda Bay, is the main source of freshwater. According to the state's freshwater inflow study for the estuary, the Colorado would provide about half the optimum flow; the Lavaca River, 17 percent; and the remainder, from other streams and runoff. The estuary is protected by Matagorda Peninsula, which stretches across the Gulf frontage and is severed by only a few passes, some natural and some man-made. Only a few miles to the southwest of the peninsula lies Matagorda Island, which protects the next system to the south—the Espiritu Santo–San Antonio Bay system.

MANAGING THE ESTUARY Two managing entities have had major effects on the shape and ecology of Matagorda Bay in the nineteenth and twentieth centuries—the Corps of Engineers and the Lower Colorado River Authority. They have wrestled with natural forces such as logjams, droughts, and floods that render use of the river very challenging. The Corps and Mother Nature have each been modifying the mouth of the Colorado River since the late nineteenth century when a channel was dredged around a 7-mile-long logjam. After the new channel became

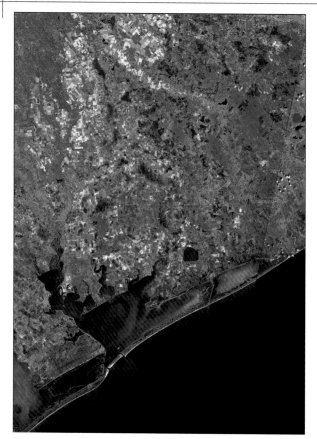

The Colorado River is the main freshwater source to the Lavaca-Colorado Estuary. Photo courtesy of the Texas Natural Resources Information System.

blocked with logs, it was cleared in 1929. The logs ended up in Matagorda Bay, where they blocked navigation and formed a new delta. In 1934 a channel was cut through the delta all the way to the Gulf of Mexico. At that point the bay was cut in two, isolating what is now East Matagorda Bay, directing the flow of the Colorado directly into the Gulf, and removing the main source of freshwater nourishing the bay. As part of a navigation project in 1968, the Corps was directed to redivert the Colorado

River back into West Matagorda Bay to enhance fisheries and to address other issues. Although diversion was completed by 1992, East Matagorda Bay still does not receive flows from the Colorado River as it did a hundred years ago.

The LCRA essentially determines the health of the Matagorda Bay system by controlling the flow of the Colorado River. Through management of the Highland Lakes, the LCRA apportions flows for municipalities in the rapidly growing Central Texas region, for recreational use of the six lakes, for rice farming near the coast, and for increasing demand from other neighboring entities outside the Colorado basin, including a proposed water supply project for San Antonio. In addition, the LCRA has to provide adequate freshwater inflows to Matagorda Bay in conjunction with the Lavaca-Navidad River Authority, which controls the Lavaca River.

The analysis of freshwater inflow needs for Matagorda Bay, published in 1997, was one of the first bay analyses completed for the state. Today the LCRA is revisiting recommendations for Matagorda Bay and management of the entire Colorado River system in its efforts to provide water to San Antonio.

The minimum recommendations for Matagorda Bay are the subject of much debate. Proposed off-channel reservoirs near the mouth of the Colorado to store water diverted from the river could reduce flows even more. Some of this stored water would be transferred out of the basin for San Antonio. There is concern that this project might alter the timing of flows to the estuary. In 2004 the LCRA began a six-year study to determine whether its project on the Lower Colorado, which includes sending water to San Antonio, will benefit the two planning regions, K and L, without harming the Colorado River or Matagorda Bay. The good news is that the state will adopt the results of the SB 3 environmental flow needs studies as official set-aside flow regimes for this and other basins. Unfortunately, the first set of recommended flows will not become official until sometime in 2010.

AREA ECONOMY AND WATER QUALITY Today Matagorda Bay is a popular recreational fishing area and has one of the state's largest fleets of shrimp boats. The combined economic impact from recreational and commercial fishing was $178 million in 1998. The bay is almost completely enclosed, except for Pass Cavallo and the man-made Matagorda Ship Channel, both of which are at the southern end. The ship channel leads north to Port Lavaca, the largest city on the bay (pop. 12,000). The ship channel, although not having close to the volume of the ports of Houston or Corpus Christi, ranks thirty-sixth in the nation in tons of foreign trade.

Lavaca Bay, the northwestern arm of the Lavaca-Colorado Estuary, receives its flows from the Lavaca River and is the center of shipping and industry in the bay. It is also the area of the bay with the most contamination problems. For years the Alcoa plant at Point Comfort discharged mercury into the bay as waste from its chlor-alkali processing operation, which has led to its becoming an underwater Superfund site. As a result, oyster harvesting in portions of Lavaca Bay have been closed for years. Fishing for finfish and crabs has been prohibited since 1988 in Lavaca Bay and nearby Cox Bay.

HISTORY OF MATAGORDA BAY Matagorda Bay is best known historically as the site where La Salle and about 180 colonists landed. La Salle thought he was landing at the Mississippi River mouth. After establishing Fort St. Louis on the Lavaca River, he explored southeastern Texas. He was killed in an insurrection by his own soldiers as he attempted to travel by land to French settlements in the upper Mississippi. In 1688 the colony succumbed to malnutrition and the Karankawa. La Salle caught the attention of the Spanish, who had practically ignored Texas since their early explorations in the 1500s. The result was increased Spanish exploration and settlement in Texas, including the creation of missions and, ultimately, the arrival of Austin and the beginning of Anglo settlement. Interest in La Salle was rekindled when the wreck of one of his ships, *La Belle*, was discovered in south-central Matagorda Bay in 1995. A cofferdam was constructed and the ship removed; it will be reconstructed at the Texas History Museum at a cost of $1.5 million.

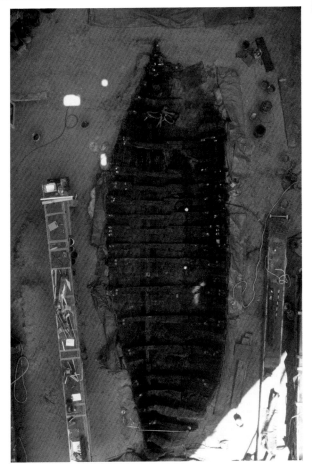

The excavation of La Salle's ship in Matagorda Bay. Photo by Wyman Meinzer.

In 1846 German colonists established a port at Indianola in West Matagorda Bay as a jumping-off point for thousands of German immigrants who settled from the coast and inland along the Colorado River to Central Texas. Indianola was also the eastern end of the Chihuahua Trail, which led to San Antonio, Austin, and Chihuahua, Mexico. A New York–based steamship line selected Indianola as its terminus, and the region soon grew to a population of five thousand. A devastating hurricane in 1875 almost destroyed the fledgling port, but it was rebuilt on a smaller scale. In 1886 another hurricane struck, and within a year the town was abandoned. Today all that remains of what was once a gateway for German immigration to Texas is a graveyard on a windswept clearing near the bay.

Guadalupe Estuary

At the mouth of the Guadalupe River lies the Guadalupe Estuary, which includes several components: Guadalupe Bay, San Antonio Bay, and Espiritu Santo Bay. San Antonio Bay is one of the least developed of all Texas bays and is home to only two towns, Seadrift (pop. 1,352) and Austwell (pop. 189). Along the northern shore near Seadrift is a chemical manufacturing plant, but the remainder of the shoreline is mainly rural, with ranches and farms. An improved channel leads from Seadrift to the Intracoastal Waterway.

Matagorda Island provides an almost continuous barrier between this bay system and the Gulf. Only the natural channels, Pass Cavallo to the north and Cedar Bayou to the south, provide interchange between bay and Gulf waters. Because Cedar Bayou often closes due to siltation, it is the subject of much controversy, and fishermen have continually lobbied to keep it open through dredging. Cedar Bayou was intentionally closed in 1979 to prevent a major oil spill originating in Mexican coastal waters from entering San Antonio Bay. Later it was reopened, but natural forces such as hurricanes alternately open and close it with regularity and cause its very location to be continually shifting.

HOME TO THE WHOOPING CRANE San Antonio Bay is best known as the location of Aransas National Wildlife Refuge and the winter home of the endangered whooping crane. These magnificent birds, affectionately known as whoopers, declined to a meager population of only 16 in the 1940s. Fortunately, the wild flock is now up to about 230 thanks to a major effort to preserve them during the last half of the twentieth century. About 50 additional cranes

Fed by the Guadalupe River, the Guadalupe Estuary includes San Antonio Bay. Photo courtesy of the Texas Natural Resources Information System.

are in a reintroduced, nonmigrating flock in Florida. More than 60 cranes are part of another migrating flock that was trained to migrate by following ultralite aircraft from Wisconsin to Florida. There is constant concern that a spill from barges on the Intra-

The endangered whooping cranes have made a comeback but still face dangers associated with use and development of their habitat. Photo courtesy of Texas Parks and Wildlife Department.

coastal Waterway, which passes through their habitat, will destroy it and thus threaten this icon of wildlife conservation.

Several studies are being carried out to determine the relationship between freshwater inflow and the survival of the cranes. Blue crabs constitute 90 percent of their diet and need certain amounts of freshwater to survive. Studies are under way to determine the amount and timing of flows needed from the Guadalupe and San Antonio Rivers into the bay to sustain the blue crab population and, as a result, the whooper population. Not surprisingly, it took a human threat to instigate the studies. A major water supply project has been conceived to take water from the saltwater barrier, 12 miles upstream of the bay, and pump it back to San Antonio and Central Texas through a pipeline about 120 miles long. Currently, this project is on hold, as its primary customer, the San Antonio Water System, no longer supports it. Scientists and stakeholders will have to consider issues such as the whooping crane habitat and the Lower Guadalupe Water Supply Project as they arrive at inflow recommendations for the Guadalupe Estuaries during the SB 3 process.

Matagorda Island is the first of a two-island chain that stretches from Pass Cavallo to Port Aransas. The island is about 50 miles long, averages 2 miles wide, and is separated by the intermittent pass, Cedar Bayou, from San Jose Island to the south. Because of its location at the southernmost end of Matagorda Bay, the island historically has served as a crucial defensive and navigational point. A lighthouse, built on the north end in 1852, has since been moved as the shoreline receded and has recently been restored. The island was the scene of frequent activity during the Civil War, and a fort was built on the northern end. During World War II, a bombing and gunnery range was built on the northern end of the island; some of the original buildings remain. Matagorda Island is uninhabited. It is now a state park and a wildlife management area, accessible only by boat. About ten thousand visitors a year enjoy its wild beaches and wetlands and ponder the eerie presence of the abandoned military buildings and airstrip. Sadly, the Texas Parks and Wildlife Department has recently closed the state park operation due to lack of funds. As a result, public access to one of the state's most magnificent natural places has been severely curtailed.

AREA ECONOMY San Antonio Bay is not well known in Texas because of its lack of population centers, ports, industrial facilities, or resorts. In spite of this anonymity, the sleepy bay plays a major role in Texas' commercial and recreational fishing. About $30 million worth of shrimp, or 13 percent of the Texas shrimp

The Matagorda Lighthouse was built in 1852 and restored in 2004. Photo courtesy of Texas Parks and Wildlife Department.

harvest, come from San Antonio Bay. In addition, 17 percent of the oyster harvest, 20 percent of the commercial crab harvest, and 500,000 hours a year of recreational fishing result in an economic value of $55 million per year. Birding and wildlife viewing also add to this total as birders from all over the world come to the area to see the whooping cranes. Although many of the charter boats to view the whooping cranes leave from the next bay to the south, Mission-Aransas Bay, the economic effects spread into the San Antonio Bay area.

A few areas of San Antonio Bay experience water quality problems due to bacterial contamination in some oyster areas and low dissolved oxygen below the saltwater barrier near the mouth of the Guadalupe River. Nevertheless, San Antonio Bay is a relatively undeveloped Texas bay that is almost completely blocked from the sea by barrier islands and, except for the Intracoastal Canal, is an isolated estuarine wilderness.

The Mission-Aransas Estuary

The Mission-Aransas Estuary comprises a series of bays that lie between San Antonio Bay and Corpus Christi Bay, including Mission Bay, Copano Bay, and Aransas Bay. Unlike most of the estuaries in Texas, Mission-Aransas has reduced river inflow, so the influence of freshwater is minimal. The 40-mile-long Aransas River and 24-mile-long Mission River are the only rivers that flow into Copano and Mission Bays, respectively, and provide freshwater for the Aransas Bay complex. During the drought of record, groundwater seeps and springs on the Mission and Aransas Rivers were the only freshwater inputs to the Aransas Bay system. The annual average of freshwater inflows amount to only 60 to 70 percent of the volume of the estuary compared to the Sabine Estuary, which receives more than fifty times its volume per year.

Like so many Texas estuaries, the Mission-Aransas is almost completely separated from the Gulf of Mexico by barrier islands, except here the separation is more notable due to the minimal natural freshwater entering the bay. The only two passes in San Jose Island, the barrier island protecting Mesquite and then Aransas Bays, are the intermittent Cedar Bayou and Aransas Pass, which connects the Gulf with the Port of Corpus Christi.

With the complex set of small side bays, barrier islands, intermittent natural passes, ship channels, and the Intracoastal Canal that connects Matagorda, Espiritu Santo–San Antonio, and Copano-Aransas Bays, the movement of freshwater and salt water is not just a simple matter of "water in–water out." Water from major rivers such as the Guadalupe will move sideways, parallel to the coast, and flow into Aransas Bay to the south and Matagorda Bay to the north. This exchange can occur in the opposite direction depending on factors such as low river flows, tidal cycles, wind, and periodic opening of new and existing cuts in the barrier islands. As water becomes more scarce and freshwater inflows are further reduced, the understanding of the movement of water between estuaries will become more important. Several state water agencies are studying these complex flows and their environmental effects.

The Mission-Aransas Estuary is not highly developed industrially, but quaint tourist businesses occupy much of its shoreline, including Mustang Island, which is the barrier island to the south side of the Aransas Pass channel. San Jose Island to the

The Mission-Aransas Estuary receives freshwater from the Aransas and Mission Rivers. Photo courtesy of the Texas Natural Resources Information System.

north of the channel is privately owned and basically uninhabited but is accessible for beachcombing and fishing on the south end for those who have access by boat. The jetties that protect the Aransas Pass extend over a mile into the Gulf and are heavily used by fishermen. Eighteen-mile-long Mustang Island is a growing tourist destination gradually expanding southward toward Corpus Christi, where a causeway links the island to the mainland. Port Aransas is located at the north end of Mustang Island, although it is not much of a port. It is mainly a fishing and tourist town and is home to the University of Texas Marine Science Institute. Port Aransas is also the entranceway to the Port of Corpus Christi. Access to Port Aransas is through a delightful short ferry ride over the Corpus Christi Ship Channel or a 20-mile drive from Corpus Christi to the south and across the causeway. Water quality issues in the area are mainly related to outbreaks of red tide that seem to plague many of the Texas bays. The causes of red tide have not yet been determined but can result in major fish kills.

To Texans, Port Aransas means fishing. From its small harbor numerous fishing guides offer excursions that may involve two-person trips or boats carrying more than fifty fishermen on multiday deep-sea trips far out into the Gulf. One of the largest Texas coastal fishing tournaments is held here. This small island community of about twenty-five hundred was once called Tarpon, referring to the world-famous sport fish that was the town's claim to fame. Tarpon fishing in the bay has declined drastically to almost nil in the Aransas area, although in recent years more

Walls in the lobby of the Tarpon Inn are covered in tarpon scales that have been autographed by the fishermen who caught them. Photo courtesy of www.thetarponinn.com.

tarpon are being caught offshore. Scientists are not certain what caused this decline. The Tarpon Inn, operating since the 1880s, stands as a memorial to that fishery. On the walls are large scales of tarpon, each two inches long or more, signed by the person who caught them. Reportedly, a scale signed by Franklin D. Roosevelt is among them.

Fishing is not the only tourist activity in Redfish and Aransas Bays near Port Aransas. Birding and wildlife viewing are thriving, and it is common to see kayaks in the bays carrying both bird-watchers and anglers. Even the wastewater plant in Port Aransas is used to supply recycled freshwater to an island marsh with board-walks known as the Port Aransas Birding Center. On the north end of Aransas Bay, the small towns of Rockport and Fulton also cater to ecotourists. Birding boat tours to the wintering home of the whooping cranes leave from these towns. Whooper tourism alone brings in $6 million each year to the local economy.

The relatively undeveloped condition of the Redfish and Aransas Bay area and its variety of marshes, coastal plain habi-tat, and sea grass beds has resulted in a portion of the bay being selected as a National Estuarine Research Reserve. This coop-erative state-national program seeks to promote stewardship of estuaries through science and education performed in protected areas. The University of Texas Marine Science Institute at Port Aransas is a key partner in this project and will administer much of the research. The selection of the Mission-Aransas Estuary is testimony to the relatively natural state of portions of Redfish and Aransas Bays.

The Nueces Estuary

Nueces and Corpus Christi Bays, at the mouth of the Nueces River, form the Nueces Estuary, which faces numerous chal-lenges from an adjacent growing city. Corpus Christi is a major port with heavy industry, a semiarid climate, limited exchange of seawater, and a significant tourism industry. In part because of these challenges, this system was selected in 1992 as one of twenty-eight estuaries designated as Estuaries of National Sig-nificance under the National Estuary Program administered

The Nueces Estuary includes Nueces Bay and Corpus Christi Bay. Photo courtesy of the Texas Natural Resources Information System.

by the U.S. Environmental Protection Agency (EPA). The only other Texas resource in the program is Galveston Bay, which was selected in 1988. The NEP designation provides federal funds to assist in the analysis of the estuary and creation of a long-term management plan. The program is titled the Coastal Bend Bays and Estuaries Program and includes not only Nueces and Corpus Christi Bays but also Copano-Aransas Bays to the north and the Upper Laguna Madre and Baffin Bay to the south.

Corpus Christi surrounds much of Corpus Christi and Nueces Bays at the mouth of the Nueces River. With a population of 277,000, Corpus Christi supports the nation's sixth largest port and the third largest refinery and petrochemical complex. Access to the port is through a 21-mile-long channel from the pass at Port Aransas that was completed in 1926. Corpus Christi, as the largest city in the Coastal Bend region, receives residual economic benefit from recreation in nearby Redfish Bay, Aransas Bay, and Laguna Madre, as well as in Nueces and Corpus Christi Bays. Tourism brings over $500 million to the local economy annually.

Although Corpus Christi Bay has one of the few natural harbors on the Texas coast, it did not become a community until 1845, when General Zachary Taylor landed his troops there in the run-up to the Mexican War. The harbor could not handle large ships, so cargo had to be loaded onto small boats for transfer. It was more than seventy-five years before the port and deepwater channel were built.

Inevitably, there are water quality problems because of the industry, shipping, and agricultural production in the area. Although water quality has significantly improved in the past twenty-five years, there is still evidence of toxic heavy metals, including lead and zinc, and an unexplainable high level of silver. As in other Texas bays, red tide is a periodic problem.

All water quality and quantity issues in Corpus Christi Bay center on the lack of freshwater inflow. The average area rainfall is 24 to 36 inches annually, while the surface evaporation is 60 inches a year, resulting in a net loss. The bay takes almost fifty months to replace its water—a long time compared to most estuaries. The Nueces River, the only major river system in the lower watershed, provides virtually all freshwater inflow to the bay. Two reservoirs, Lake Corpus Christi and Choke Canyon Reservoir, built on the Nueces and Frio Rivers, have seriously affected flows to the bays. When the Choke Canyon Reservoir was built upstream by the city of Corpus Christi in the early 1980s, minimum

Corpus Christi is the largest city in the Coastal Bend region. Photo courtesy of the Corpus Christi Convention and Visitors Bureau.

flows were mandated to be released for estuarine maintenance, but the flows appear to be insufficient. The Texas Shrimp Association filed suit in 1990 due to the lack of freshwater coming into the bay. It took several years for the Choke Canyon Reservoir to fill up because of a drought at the time, and subsequently the bay received little freshwater inflow. Contributing to this problem is the continuing drought in parts of South Texas. The bottom line is that river flows since construction of the Choke Canyon dam are 28 percent lower annually than before the dam was built.

The reduction of freshwater inflow has been partially responsible for the decline of the oyster population in the bay, which once had the largest oyster reefs in Texas. The other contributing factor to this decline was oyster shell mining in the bay, which continued as late as the 1970s, to support the construction industry and the manufacture of magnesium. These same factors are also partially responsible for the reduction in the number of islands in the bay. As part of the Coastal Bend Estuary Management Plan, lost island habitat will be restored, providing a beneficial use for dredge spoil and creating new areas for bird nesting.

Because of the configuration of the Nueces delta, freshwater from the Nueces River can only reach the delta wetlands at higher flows than now normally occur. The base flow of the Nueces River entered a part of the bay without wetland marshes that are so crucial for species in their early stages. A partnership between the city of Corpus Christi and the Nature Conservancy purchased land in the upper part of the delta where an overflow channel was built. Today freshwater can be delivered to the interior delta wetlands at lower flows via this channel. In addition, the city has rerouted some treated return wastewater to the delta wetlands as supplemental flow.

The south end of the barrier island, Mustang Island, is connected to Corpus Christi by the Padre Island Causeway, providing a highway connection to Port Aransas. The causeway also connects to the north end of the longest undeveloped barrier island in the world—Padre Island. Some critics say the causeway has restricted flows into the Laguna Madre, increasing its salinity, which could encourage outbreaks of brown tide. The height of the causeway has also been an issue, as it was constructed but a few feet above sea level. This was a particular concern because the causeway is the escape route from Padre Island and southern Mustang Island in the event of a hurricane. In response to these concerns, the elevation of the causeway was completed in 2005, creating greater security for evacuation and increased interchange of water below the bridge.

Corpus Christi Bay, like Galveston Bay, faces tremendous stress to its ecosystem from industry, navigation, and recreation. The good news is that with efforts that include the National Estuary Program plans are under way to manage and sustain estuarine habitat. The major challenge for Corpus Christi Bay remains finding enough freshwater to sustain a healthy estuary and, at the same time, meet growing municipal and industrial needs. It is possible that desalination of salt water is one answer, and Corpus Christi seems to be an ideal location, if the brine residue can be disposed of properly. Even with the SB 3 mandated studies and freshwater inflow recommendations, there may not be enough available water in this basin to set aside for environmental flows.

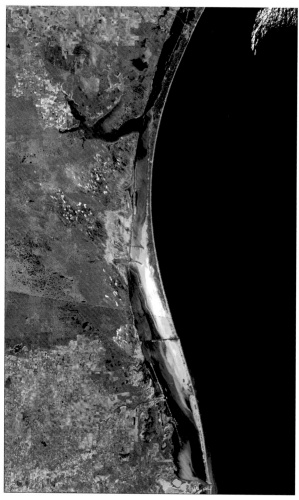

The Laguna Madre is one of five hypersaline lagoons in the world. Photo courtesy of the Texas Natural Resources Information System.

The Laguna Madre

Padre Island runs south from Corpus Christi for 130 miles, forming an almost continuous barrier between the Gulf of Mexico and the Laguna Madre and effectively creating an enclosed

Newly hatched Kemp's Ridley sea turtle makes its way to the water. Photo courtesy of Texas Parks and Wildlife Department.

hypersaline lagoon for the length of the island. The island has the longest sand beach in the United States, and much of it is in Padre Island National Seashore, managed by the National Park Service. This beach is the primary nesting ground in Texas for the endangered Kemp's Ridley sea turtle. The island and the mainland shoreline of Laguna Madre are still mostly uninhabited. Historically, there was little settlement or activity in and around Laguna Madre. One of the few events on the island occurred during the Spanish era in 1854 when four ships laden with silver wrecked in the general area of what is now Port Mansfield. Only one ship survived, and, amazingly, much of the silver was recovered within a year of the wrecks. The wrecks of the three ships were discovered in the 1960s and 1970s and the artifacts are now in possession of the state. During the Civil War, Confederate blockade runners used Laguna Madre as a connection to an overland route to the cotton markets of Mexico. Since the 1850s most of the mainland coastline is still occupied by large ranches, including the fabled King Ranch.

The only pass through Padre Island to the Gulf is at Port Mansfield about two-thirds of the way down the island. This channel, dug in 1962 to connect the Gulf to the Lower Laguna Madre, allows tidal exchange between the Gulf and Lower Laguna Madre, which had been almost nonexistent in its natural state. As a result of the pass and the Intracoastal Waterway, there are more abundant populations of finfish such as redfish, flounder, and speckled trout. In fact, Laguna Madre has become a world-famous recreational fishery. Oysters are not abundant in the Laguna Madre due to its high natural salinity. Some types of shrimp are present, and the commercial brown shrimp industry is now quite successful. This is one of only five hypersaline lagoons in the world and the only one in the United States. In some areas, especially in the Upper Laguna, the Laguna Madre can be even saltier than the sea. The high salinities and sandy, shelly Padre Island sediments as well as sandy soil on the inland shore make for extremely clear water. Much of the Laguna Madre is less than a foot deep, and the average depth is about 3 feet.

SEA GRASSES IN THE LAGUNA The key habitat for the success of finfish is the abundance of marine sea grasses, and Laguna Madre contains about 75 percent of the sea grass in Texas. The redhead duck is also highly dependent on the most common type of Laguna sea grass, called shoalgrass. More than 75 percent of the world's population of redhead ducks winters in the Laguna, including the Mexican portion. South of the Rio Grande Delta in Mexico is the Laguna Madre de Tamaulipas, which has characteristics similar to its northern counterpart. Several joint projects between the Nature Conservancy and Mexico have been conceived to protect the Mexican Laguna Madre. These

Sea grasses are among the most biologically, recreationally, and economically valuable habitats in the world. They provide vital feeding and nursery habitat for waterfowl, fish, shrimp, crabs, and countless other estuarine species. Photo courtesy of Texas Parks and Wildlife Department.

efforts successfully thwarted plans for a Cancun-style resort and an extension of the Intracoastal Waterway through the Mexico Laguna Madre. However, discussions of the proposed development plan have resurfaced recently.

The sea grass meadows of the Laguna Madre are undergoing change, with the Upper Laguna Madre showing an increase in seagrass meadows and the Lower Laguna showing a decrease. Some of this change in deeper water areas of the Lower Laguna has been attributed to reduction of light due to turbidity from open-bay disposal of dredge spoils from the Intracoastal Waterway. Also, careless recreational boating in this shallow estuary can scar sea grass beds. The increase of sea grass in the Upper Laguna can be attributed to more moderate salinities resulting from the Intracoastal Waterway, allowing better water circulation; however, this increase is offset to some extent by the effects of the brown tide, a microscopic algae that blocks sunlight. There have been persistent blooms of brown tide in the Upper Laguna Madre intermittently since 1990. These factors have altered the sea grass varieties, especially the shoalgrass on which the redhead duck depends. Thus the continued health of the Laguna Madre de Tamaulipas is of increasing importance as a reserve feeding area, as the northern Laguna Madre struggles with drought or an outbreak of red or brown tide.

Solutions to sea grass reduction in the Laguna Madre include a consideration by the Corps of Engineers of alternative disposal methods for dredged material to reduce turbidity problems. The reasons for the more recent outbreaks of red and brown tides are not fully understood. However, scientists speculate that the increase in harmful outbreaks may be related to increased nutrients in nonpoint source pollution from development on the mainland and from agricultural runoff.

WATER QUALITY IN THE LAGUNA The Lower Laguna Madre receives freshwater flow from the Arroyo Colorado, an old channel of the Rio Grande that has been modified to perform a number of tasks. Irrigation water for farms is delivered through the Arroyo, and some of its chemical laden residue is then returned to the stream. Many municipalities in the Lower Rio Grande Valley discharge wastewater into the Arroyo, and floodwaters from the Rio Grande are diverted into it to prevent flooding in Brownsville. Barges travel inland to Harlingen on the old channel. There are several water quality issues on the Arroyo that have arisen from these activities, and fish consumption advisories have been posted from time to time. The Lower Laguna Madre receives much of this effluent, although data do not indicate the presence of toxic chemicals in the Arroyo.

Texas leads the nation in shrimp farming, producing 70 to 80 percent of the U.S. output. The presence of this activity adjacent to the Arroyo is another source of pollution to the Lower Laguna Madre. These discharges may contain pesticides, antibiotics, and suspended organic matter. An additional risk from shrimp farming is the possible escape of non-native or so-called exotic species of shrimp. Exotics have been found in the bay, and scientists are worried that they may become dominant or spread a disease to the native species for which they have no im-

South Padre Island is a popular tourist destination. Photo courtesy of Texas Parks and Wildlife Department.

munity. As Texas also produces about 25 percent of the nation's wild shrimp catch, measures have been taken to ensure the survival of these two industries and to protect the environment on which they depend.

The Laguna Madre is probably the most distinctive and remote of Texas estuaries. It is almost completely landlocked, generally unaffected by tides, extremely shallow, and lacks perennial streams. As such, the Laguna Madre is highly vulnerable to human activities that can cause serious environmental imbalances. Care needs to be taken to preserve this fascinating body of water.

SOUTH PADRE ISLAND At the southern end of Padre Island lies the city of South Padre Island (or simply South Padre), an international resort community on a very narrow strip of land that boasts high-rise hotels of more than four thousand rooms. Not only is the city the spring break destination for thousands of U.S. college students, but the island attracts a large contingent of tourists from Mexico, the United States, and elsewhere. Connected to the mainland by the Queen Isabella causeway, this thriving resort is highly vulnerable to natural and man-made forces. For example, one night in 2001 a barge tore through the causeway. Several people died as they drove off the causeway into the dark lagoon. Ferry boats from Port Aransas over 100 miles away were sent as emergency transport for those stranded on the island. The causeway was reconstructed in three months. This threat to Padre Island pales in comparison to the inevitable threat by seasonal hurricanes. Most of the existing buildings have been built since the last hurricane in 1966, which severely damaged the town and washed more than thirty new channels through the island.

Rio Grande Inflows

The southern tip of Texas is delineated by the sometimes closed, nonflowing mouth of the Rio Grande (Mexico's Rio Bravo). When flowing, the river empties directly into the Gulf unimpeded by barrier islands or an estuary. In recent years, drought in Mexico and larger withdrawals from the Rio Conchos have completely stopped the river from reaching the Gulf. Thus this longest of Texas' great rivers is the poster child for the need to protect freshwater inflows. Though there still remain miles of spectacular estuaries and undeveloped beaches and shorelines, as the population grows and major water development projects come on line, the remaining natural areas of the Texas coast will face enormous pressure.

Honey Creek State Natural Area is adjacent to Guadalupe State Park. Photo courtesy of Texas Parks and Wildlife Department.

5. WHO'S WHO IN WATER

Numerous government institutions at the local, state, and federal levels share responsibility for managing Texas' water. A handful of these agencies have the most control. The key agencies that influence water use in Texas are the Texas Commission on Environmental Quality, the Texas Water Development Board, the Texas Parks and Wildlife Department, the U.S. Army Corps of Engineers, and the U.S. Environmental Protection Agency. A number of river authorities also exert influence on water use and regulation. Recently, a growing number of groundwater conservation districts have become more of a force in water resources discussions. There are also nonprofit organizations, such as the Texas Water Conservation Association, whose members include municipalities and river authorities that wield heavy influence in the water arena. Many other nonprofits, such as the Sierra Club, the National Wildlife Federation, and Environmental Defense, advocate for an environmental perspective on water management. The following is a brief summary of the major agencies, government entities, research institutions, and nonprofit groups that influence water policy in Texas.

STATE AGENCIES AND ENTITIES

Texas Commission on Environmental Quality (TCEQ)

The TCEQ oversees the granting of water rights, the issuance of wastewater permits, and matters related to water quality and drinking water systems. More than any other state or federal agency, the TCEQ affects water quantity and quality in Texas. Seated at the head of this large and powerful institution are three commissioners appointed by the governor. Many of the duties and functions of the TCEQ involving the preservation of water quality are mandated by federal laws and rules of the EPA that have been delegated to the state for implementation.

STATE WATERMASTER PROGRAM The TCEQ is responsible for administering the State Watermaster Program. In many of the river basins in Texas, compliance by water rights holders is on the honor system. However, in three areas of the state there is a designated watermaster to ensure compliance with water rights. Watermasters monitor water use, including diversions, in-stream flow rates, and reservoir levels. There are two state watermaster programs: the Rio Grande Program and the South Texas Program. The Rio Grande watermaster controls the Rio Grande Basin from Fort Quitman to the coast. The South Texas watermaster's jurisdiction comprises seven watersheds in fifty counties in south-central Texas and was recently expanded to include the Concho River Basin, a tributary of the Colorado River.

Texas General Land Office (GLO)

The GLO manages the public lands of Texas, including the coastal submerged lands 10.3 miles into the Gulf of Mexico. Its responsibilities include management of mineral rights, oil and gas leasing, and riverbeds. Recently, the GLO began leasing water rights under Texas government lands to municipalities and others. This began in the 1950s when the University of Texas leased rights to six West Texas cities, including Midland and

Estero Llano Grande State Park is a major birding area of Texas. Photo courtesy of Texas Parks and Wildlife Department.

Odessa. The GLO signed a lease of water rights with the municipality of Presidio in 1993. In fall 2006 the GLO officially began accepting applications for the leasing of state lands to private investors for the purpose of extracting groundwater to sell to municipalities. This policy change has been controversial and thus far unsuccessful, particularly as the lands involved are located in Texas' most arid regions.

Texas Parks and Wildlife Department (TPWD)

TPWD is the agency responsible for protecting the state's fish and wildlife resources as well as administering its parklands and wildlife areas. The agency controls all aspects of hunting and fishing, including issuance of licenses. Its mission includes management of coastal biological resources and the protection of the fish and wildlife species that inhabit them. Its nine-member board is also appointed by the governor. TPWD, by state and federal law, may comment on water rights, discharge, and wetland alteration applications with respect to their effects on wildlife and plants, both inland and in bays and estuaries. The department is advised by the Texas River Conservation Advisory Board, the Freshwater Fish Advisory Board, the Coastal Fisheries Advisory Board, and other entities on water-related topics.

Texas Water Development Board (TWDB)

The TWDB is the principal state agency for water planning and financial assistance for water development projects. Like TCEQ and the Texas Parks and Wildlife Commission, the TWDB, which consists of six members, is appointed by the governor.

The TWDB publishes a state water plan every five years. This document has significant influence on water project funding and legislation. The board's assessment process entails scientific modeling and analysis of surface and groundwater availability in Texas. The board also has influence in the water rights process, as any new water rights application must be consistent with the water plan.

The original purpose of the TWDB was to provide low-interest loans to poor communities for water improvement projects. Today the TWDB provides these loans to all levels of governments within the state, as well as to nonprofit water utility companies.

In addition, the TWDB administers the Water Bank to facilitate temporary or permanent water rights transfers between sellers and buyers, and the Water Trust, where water rights can be donated, leased, or purchased to provide instream flows for environmental purposes. There are currently only three deposits in the Water Trust: one deposit of 33,000 acre-feet from the San Marcos River donated by Texas State University and two deposits totaling approximately 1,200 acre-feet from the Rio Grande donated by private citizens.

Texas Railroad Commission

The Railroad Commission, in addition to the management of railroads, regulates the oil and gas industry. It affects water quality aspects of oil and gas production, such as deep well brine injection and many other processes associated with petroleum pumping and refining.

Regional Water Planning Groups

Landmark legislation, known as Senate Bill 1, provided that under the guidance of the TWDB there be sixteen water planning regions in Texas, each with a planning group composed of a minimum of eleven members. The members of the group are to include representatives of the following interests:

- General public
- Counties
- Municipalities
- Industry
- Agriculture
- Environmental

Sixteen regional water planning groups help manage existing water resources and plan for future needs on a geographic scale. Data Source: Texas Water Development Board.

- Small business
- Electrical generation
- River authorities
- Water and groundwater districts
- Water utilities

Elements of the regional plans are area description and present and future population statistics, water demand projections, supply management strategies, recommendations and impacts, and funding recommendations to the legislature. All new water rights applications must be part of this planning process. Plans were completed in 2001 and 2006, with public hearings held prior to approval and publication. Regional water plans are incorporated into the TWDB's state water plan, which is completed within a year of the regional plans. In the short history of this process, its bottom-up system has brought many improvements to water planning. However, a general lack of consideration for environmental concerns, a lack of watershed coordination, and regional conflicts have caused problems and exposed shortcomings in the system.

River Authorities

River authorities are quasi-governmental agencies that are generally responsible for managing and developing the water resources of their respective basins. All of the river authorities in Texas have boards appointed by the governor, with the exception of the San Antonio River Authority, which holds elections every two years. They do not have taxing authority, and their revenue mainly comes from water and power sales and wastewater plant projects. Many river authorities are also involved in water quality issues, conservation programs, promotion of water recreation, park management, and educational programs. River authorities can buy and sell water as well as manage reservoirs, often in conjunction with the U.S. Army Corps of Engineers or the U.S. Bureau of Reclamation. Following is a list of the river authorities of Texas:

- Angelina-Neches River Authority
- Brazos River Authority
- Guadalupe-Blanco River Authority
- Lavaca-Navidad River Authority
- Lower Colorado River Authority
- Lower Neches Valley Authority
- Nueces River Authority
- Red River Authority
- Sabine River Authority
- San Antonio River Authority
- San Jacinto River Authority
- Trinity River Authority
- Upper Colorado River Authority
- Upper Guadalupe River Authority

The river authority system in Texas has often been frustrating to interests as diverse as governors and environmentalists, mostly due to the perceived lack of accountability in its semiautonomous structure.

Groundwater Districts

In Texas groundwater accounts for 59 percent of all water consumed statewide, but its extraction and use are virtually unregulated. In recent years, eighty-four local groundwater districts covering most of the state's aquifers and enabled by another landmark law, Senate Bill 2, provide various levels of regulation. Unfortunately, there are still some areas of the state with no groundwater districts, allowing landowners to pump unlimited amounts without regard to the adjacent wells. Some districts, like the Edwards Aquifer Authority, cover the hydrologic boundaries of the aquifer, but most groundwater districts are bound to a single county, though the underlying aquifer may sit under several. Finally, most districts lack adequate funding or scientific capacity to fulfill their mandate to manage groundwater.

In some cases, different aquifers overlay each other, resulting in overlapping district jurisdiction. For instance, in Hays County, the Edwards Aquifer Authority (EAA) manages the Edwards Aquifer, but in parts of the county the Trinity Aquifer lies below

The Sabine River meanders through East Texas. Photo courtesy of Texas Parks and Wildlife Department.

the Edwards. Therefore, the Hays Trinity Groundwater Conservation District was created to manage the Trinity. In some counties over the Edwards Aquifer, however, the Trinity is still not governed by a groundwater district. This situation has allowed landowners to drill through the Edwards to pump from the Trinity and avoid the permitting process required by the EAA.

The EAA, though arguably the most well-funded district in

There are eighty-nine Groundwater Conservation Districts in Texas, eighty-four of which are confirmed. Data source: Texas Water Development Board.

the state, has historically been caught between pumping interests over the aquifer, principally San Antonio and downstream interests from San Marcos and New Braunfels to Seguin and Victoria, which lack representation on the Authority Board.

Soil and Water Conservation Districts

Soil and Water Conservation Districts were established to promote and facilitate conservation of soil and water resources in ranching and farming areas of the state. Directors of the 217 districts are elected at annual conventions and must be agricultural landowners. The statewide Soil and Water Conservation Board oversees the districts by providing financial support and communicating with the legislature. In addition, the board administers the state's soil and water conservation law and coordinates nonpoint source pollution reduction and conservation efforts. Both of these entities work closely with the National Resources Conservation Service (NRCS), under the U.S. Department of

Agriculture, which provides technical and financial assistance for conservation activities. The districts and the State Board work in partnerships with other local, state, and federal agencies as well. Ongoing partnership projects include analysis of agricultural impacts on the Bosque River west of Waco and various water quality impacts on Plum Creek, a tributary of the San Marcos River.

Local Water Entities

There are numerous entities that regulate water on a smaller scale than regional or state organizations. Most of us are familiar with city-owned water supplies, but often there are water suppliers that operate either on a larger scale than just a municipality or in areas of a county not covered by a city water supply. Some local water entities may also operate wastewater plants. The Texas legislature authorized the creation of these entities, and they were generally created by local petitions or requests from county commissioners; however, the mechanisms for each are different. In addition to water supply organizations, there are entities that in some manner affect water, waterways, or the use of water. For example, levee and navigation districts also exist in some areas of Texas. Following is a list of some of the organizations that have influence over water and waterways in Texas on a local level:

- Municipal Water Supply Districts
- Water Improvement Districts
- Water Control and Improvement Districts
- Fresh Water Supply Districts
- Drainage Districts
- Levee Improvement Districts
- Navigation Districts
- Municipal Utility Districts

FEDERAL AGENCIES AND ENTITIES
U.S. Army Corps of Engineers (Corps)

The Corps' mission is to plan, design, build, and operate wa-

Navarro Mills Lake, east of Waco, is one of many reservoirs managed by the Army Corps of Engineers. Photo courtesy of the Trinity River Authority.

ter resource and other civil works projects to facilitate navigation, flood control, and environmental protection. Much of the Corps' presence in Texas involves the construction and operation of major reservoirs on the Gulf Intracoastal Waterway and navigation channels at every port on the coast. The Intracoastal Waterway is an inland waterway composed of a series of constructed channels and improved riverways stretching from Florida to Brownsville, Texas. The Corps also issues what are called 404 permits that enable construction in wetland areas.

U.S. Department of Agriculture (USDA)

NATURAL RESOURCE CONSERVATION SERVICE (NRCS) This arm of the USDA provides financial and technical assistance to farmers and ranchers in soil and water conservation projects on their land. Flood control dams on creeks to protect small towns are also planned and facilitated under the NRCS. The cities of San Marcos and Sanderson have networks of these types of dams. The NRCS also distributes funding that can be used to improve watershed function on private lands through habitat management and the establishment of conservation reserves.

U.S. FOREST SERVICE The USDA oversees the nation's national forests, including 633,000 acres of nine forests, grasslands, and wilderness areas in Texas. In Texas, the Trinity, Neches, Sabine, and Angelina Rivers flow through national forests, as do numerous small streams and springs, creeks and bayous.

U.S. Department of the Interior (DOI)

NATIONAL PARK SERVICE (NPS) Texas has two national parks, Big Bend and the Guadalupe Mountains, which are administered by the NPS under the DOI. The NPS also administers the Wild and Scenic River portion of the Rio Grande adjoining the Big Bend National Park, Big Bend Ranch State Park, and the Black Gap Wildlife Management Area. Padre Island National Seashore and Big Thicket National Preserve in East Texas are also administered by the NPS.

U.S. BUREAU OF RECLAMATION The Bureau of Reclamation is mainly active in the far western United States, where it has constructed and manages reservoirs, irrigation facilities, and hydroelectric dams. In Texas, the bureau has constructed and is currently responsible for five reservoirs, including Choke Canyon Lake, as well as municipal and industrial water projects.

U.S. FISH AND WILDLIFE SERVICE (FWS) The FWS is the DOI agency responsible for conserving, protecting, and enhancing fish, wildlife, plants, and habitats for the American people. The FWS shares responsibility with the National Oceanic and Atmospheric Administration (NOAA) for administering the Endangered Species Act. The FWS manages several national wildlife refuges in Texas, including Aransas National Wildlife Refuge. The agency's impact on water resources comes mainly through enforcement of the Endangered Species Act and participation in the proceedings of the TCEQ and the Corps.

U.S. Environmental Protection Agency (EPA)

The EPA ensures that states comply with, among other things, federal water quality acts, including the Drinking Water Act and the Clean Water Act. Although Texas, like most states, has been delegated authority to issue discharge permits under the Clean Water Act, it still has to report to the EPA on implementation and the status of water quality in the state. The EPA also manages federal Superfund sites, which are locations polluted by toxic chemicals that the government has determined must be cleaned up but for which responsibility has not been fully determined and funding has not been secured.

U.S. Geological Survey (USGS)

USGS is the world's largest water, earth, and biological sciences and civilian mapping agency. It collects, monitors, analyzes,

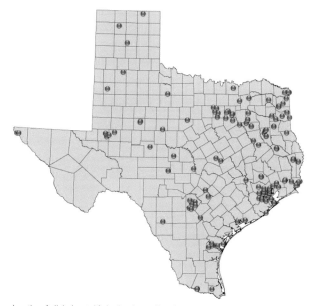

Location of all designated federal and state Superfund clean-up sites. Forty-three of the most serious sites are listed on the EPA's National Priorities List and may need long-term clean-up. Data Source: Texas Commission on Environmental Quality.

and provides scientific information about natural resources. The National Streamflow Information Program in the USGS operates and maintains about 7,300 stream gauges throughout the nation, including 400 in Texas. These gauges record water quality and quantity information in intervals around the clock. The real-time data of many of these gauges are available online. In recent years, funding for this program has declined severely, threatening water resources management in the state.

U.S. International Boundary and Water Commission (IBWC)

The IBWC is an international body that is connected to the U.S. Department of State on the U.S. side of the Rio Grande River and a sister component in Mexico. It oversees boundary and water treaties between the United States and Mexico and monitors the water quality and quantity of the Rio Grande. Since 1992 this agency has been preoccupied with issues involving shortages of water owed to the United States by Mexico. The two countries have signed two treaties apportioning the water of the Rio Grande and its tributaries, along with other agreements affecting the river channel.

Federal Emergency Management Agency (FEMA)

FEMA's water-related role is to assist state and local entities during and after flood events, such as the flooding caused by Hurricane Katrina in 2005. It also administers the National Flood Insurance Program, which determines 100- and 500-year floodplains. Determination of exact floodplain survey lines is often controversial because it affects where development can occur and in what manner. FEMA's role has been expanded since it became part of the U.S. Department of Homeland Security in 2001 to include response and recovery efforts following any national incident. This reorganization has, according to many, been responsible for wide-scale waste and service failure in the wake of Hurricane Katrina.

NONPROFIT CONSERVATION ORGANIZATIONS

Nonprofit conservation organizations can play roles that gov-

Workers from the Federal Emergency Management Agency clean up New Orleans after the Hurricane Katrina disaster in 2005. Photo by Jocelyn Augustino and courtesy of the Federal Emergency Management Agency.

ernment agencies sometimes cannot or do not fill. They can advocate for specific conservation issues that a state or federal agency might be restricted from supporting due to laws or political influence. An exhaustive list of the state and national nonprofit conservation organizations that work to protect the natural areas of Texas is not possible here. Below is a discussion of just some of the many groups that work on statewide issues or that influence the way Texas manages its natural environment.

Nonprofit Conservation Organizations in Texas

Caddo Lake Institute: The institute's mission is to protect and improve the cultural and ecological integrity of the Caddo Lake ecosystem, a wetland area in the Cypress Creek watershed of Texas and Louisiana. The 1993 Ramsar Convention named Caddo Lake a "wetland of international importance." The institute focuses its work on a data collection clearinghouse and on community education.

Galveston Bay Conservation and Preservation Association: The association formed in 1974 and successfully stopped the U.S. Army Corps of Engineers from constructing a levee. Inac-

Texas wild rice is an endangered species found in the headwaters of the San Marcos River. Photo courtesy of Texas Parks and Wildlife Department.

tive from 1992 to 1997, the group reorganized to address current issues affecting the Galveston Bay area with some success.

Greater Edwards Aquifer Alliance (GEAA): Formed in 2002, this group of citizen-based organizations ranging from Austin to Del Rio created the Edwards Aquifer Protection Plan. The plan calls for elected officials to repeal the grandfather clauses for surface water and the rule of capture for groundwater and take other steps to reduce pollution and pressure on Central Texas' water resources.

San Marcos River Foundation (SMRF): SMRF was formed in 1985 to protect water quality and quantity of the natural spring-fed river for its ecological value, endangered species, and access for human use. SMRF's most recent campaign has been to apply for water rights for instream flows in the San Marcos and Guadalupe Rivers. The initial application was denied by the TCEQ's board. SMRF, along with Caddo Lake, Matagorda Bay, and Galveston Bay conservation groups, has taken TCEQ to court over instream flows water rights; the cases are now pending.

Save Our Springs Alliance (SOS): SOS was formed in 1991 in response to increasing development pressures in Austin. Its focus has been on regulating development in the Barton Springs–Edwards Aquifer watershed to protect water quality in the aquifer, its springs, and the endangered species in the area.

Texas Center for Policy Studies (TCPS): Since 1985 the TCPS' mission has been to improve quality of life in Texas in the face of growth and development by providing information to citizens and policy makers.

Texas Conservation Alliance (TCA): Formerly the Texas Committee on Natural Resources, this organization focuses its efforts on conserving and protecting native ecosystems. The committee has provided input on state water quality standards and on proposed reservoir projects. TCONR is also the Texas affiliate of the National Wildlife Federation.

Texas Rivers Protection Association (TRPA): TRPA works to protect river flow, water quality, and natural beauty while at the same time educating the public about the importance of conservation and safe recreation on Texas' rivers. Members include representatives from landowner coalitions, conservationists, canoe clubs, and fishing associations.

Texas Water Conservation Association (TWCA): This organization is dedicated to the conservation, development, protection, and use of the state's water resources for all beneficial purposes. Its members and board of directors consist of a wide range of water users from irrigators to river authorities to environmental organizations. Its purpose is to act as an advocate for water users by advising multiple levels of government, informing the public, and providing insurance to members.

Texas Water Matters: This collaboration among Environmental Defense, National Wildlife Federation, and the Lonestar Chapter of the Sierra Club works to ensure that there is enough water for human and environmental needs, to reduce future water demand, and to educate decision makers and citizens about the importance of sustainable water resources.

Texas Wildlife Association (TWA): This association is made up of hunters, anglers, wildlife observers, and conservationists who work together with private landowners to protect habitat, wildlife, and the natural resources of the state.

Wimberley Valley Watershed Association: The association promotes sustainable watershed management to ensure water quality and quantity through educational programs. It is especially concerned with groundwater resources and their connection to water use and surface water.

National Nonprofit Conservation Organizations

American Rivers: This organization is dedicated to protecting and restoring rivers to benefit humans, fish, and wildlife. Each year the group identifies the ten most endangered rivers in the nation based on the importance of the river and the threats it faces. Since 2000 the Trinity, Rio Grande, Guadalupe, and San Jacinto Rivers have been listed among the most threatened in the United States.

Coastal Conservation Association (CCA): The mission of the CCA is to protect coastal resources for the benefit and enjoyment of the public by focusing on education and by influencing local, state, and national legislation. Created in 1977, the CCA has fifteen coastal state chapters.

Environmental Defense (ED): Founded in 1967, ED researches environmental problems and provides scientific solutions. The organization is dedicated to protecting environmental rights, including clean air, clean water, healthy food, and ecosystems.

Jacob's Well, 100 feet down. The Wimberley Valley Watershed Association is dedicated to protecting this unique spring, which is the headwaters of Cypress Creek. Photo by Ryan Eastman.

National Fish and Wildlife Foundation: This foundation was created by Congress in 1984 to generate funds for the conservation of fish, wildlife, plants, and habitats. It receives congressional funding and distributes money through a matching funds grant program for conservation purposes. Furthermore, the foundation fosters partnerships between the public and private sectors while investing in conservation practices and sustainable resources.

National Wildlife Federation (NWF): This nonprofit is committed to protecting wildlife and habitat. It has been instrumental in restoring endangered species to their original ranges, such as the return of wolves to Yellowstone National Park. In addition, the NWF monitors U.S. Army Corps of Engineer's water-related projects closely to ensure that ecological damage is kept to a minimum.

Sierra Club: With over 750,000 members, the Sierra Club is the nation's oldest and largest grassroots environmental organization. For more than one hundred years, through activism, education, and conservation programs, the Sierra Club has been instrumental in conserving wildlife, protecting over 150 million acres of wildlife habitat, and creating many national parks.

STATEWIDE STAKEHOLDER AND PROFESSIONAL ASSOCIATIONS IN TEXAS

Texas has many water-related organizations that bring together stakeholders from different regions, professions, and user needs. Some of these organizations are nonprofits, and some are quasi-governmental. The following is a list of just some of the major organizations:

- American Water Works Association—Texas Section
- Association of Texas Soil and Water Conservation Districts
- Texas Alliance of Groundwater Districts
- Texas Groundwater Association
- Texas River and Reservoir Management Society
- Texas Rural Water Association
- Texas Water Utilities Association
- Water Environment Association of Texas

WATER RESEARCH AND EDUCATION ENTITIES

There are many institutions in Texas that perform water resources–related research. Some are associated with universities; others operate independently. The following list is not intended to be comprehensive but gives an indication of the various institutions statewide.

- Bureau of Economic Geology at the University of Texas at Austin
- Center for Research in Water Resources at the University of Texas at Austin
- Center for Water Research at the University of Texas at San Antonio
- Environmental Institute of Houston at the University of Houston at Clear Lake

- Harte Research Institute for the Gulf of Mexico at Texas A&M University—Corpus Christi
- Houston Advanced Research Center
- Institute for Environmental and Human Health at Texas Tech University
- Institute for Environmental Studies at Texas Christian University
- Texas Water Resources Institute at Texas A&M University
- The River Systems Institute at Texas State University— San Marcos
- The Texas Institute for Applied Environmental Research at Tarleton State University
- University of Texas Marine Science Institute

Waterfall on the Devils River in West Texas. Photo by Wyman Meinzer.

6. TEXAS WATER LAW

A Blend of Two Cultures

Water law in Texas has evolved from the cultures and legal systems of Mexico and the eastern United States. Early Spanish settlers established irrigation-intensive settlements with systems of ditches, or acequias, that were managed by the communities they served. This centralized control of water, based in part on Spanish civil law, is the origin of Texas' system of prior appropriation. Prior appropriation refers to a system of administering water based on a government-issued permit that includes the principle of first come, first served, where the most senior water rights holder has first rights to available water. Many western states base their water law on the doctrine of prior appropriation.

Anglo-American settlers moved into Texas in the early 1800s, bringing with them the legal system of riparian law, which was based on English common law. Riparian law allows landowners bordering streams the right to use water. The riparian system, originating in England where rainfall is plentiful, proved inadequate in Texas with its scarce water supplies and frequent droughts.

Over time, riparian law was gradually merged with prior appropriation, which is the basis of current Texas water law. There are still remnants of the old riparian law in a provision that allows a 200-acre-foot exemption for domestic and livestock use for riverside landowners. Other than this domestic and livestock exemption, today all surface water is controlled by permits issued by the state.

The monumental task of merging the assorted riparian and prior appropriation water rights began with the passage of the 1967 Water Rights Adjudication Act. More than 11,500 claims for historic riparian rights were filed that had to be adjudicated. As of late 2006, all water rights claims had been adjudicated with the exception of pending claims on the Upper Rio Grande and the Pecos River.

The Mission Espada Aqueduct was built in 1745 to transport river water to irrigate adjacent fields. It is now part of the San Antonio Missions National Historical Park. Photo by Gregg Eckhardt and courtesy of www.edwardsaquifer.net.

HOW DO YOU APPLY FOR A WATER RIGHT?

Surface water in Texas is owned by the state and held in trust for everyone to use. However, citizens must apply to the Texas Commission on Environmental Quality for the right to withdraw and consume it from lakes, rivers, and streams. These requests must meet several criteria for a permit to be granted, including whether the use will be beneficial and whether enough water is available. Following is a list of the accepted beneficial uses for surface water.

- Domestic and municipal
- Industrial
- Agricultural
- Mining
- Hydroelectric power
- Navigation
- Recreation and pleasure
- Public parks
- Game preserves
- Other beneficial purposes of use recognized by law, including instream uses, water quality, aquatic and wildlife habitat, or freshwater inflows to bays and estuaries

HOW MUCH WATER IS AVAILABLE?

Determining how much water is available is often a contentious part of the permitting process because keeping track of water rights is very complicated: not all water rights are fully used, and they are not all used at the same time. For example, farmers use water at certain times of the year, whereas cities consistently use water throughout the year. Further, the presence of water flowing in a river does not necessarily mean that there is water available for the state to grant.

The flow of a river can have owners, destinations, and uses that are not obvious. In some situations, such as high flows from floods, the water is simply not available for permitting because of its unpredictable availability and amount. In other cases, flow in a river can be headed downstream to a senior water right or to a reservoir where it is committed for the storage pool to be sold to municipalities. The downstream water rights holders may not even be consuming any or all of that water at a given time, but that water is still reserved for their use. In another scenario, wastewater dischargers upstream provide flows that could cease if the discharger decided to directly reuse all of its wastewater. To complicate the matter, the discharger could attempt to divert its wastewater downstream at another point and then reuse it; this is known as indirect reuse and is hotly contested. Sometimes flow in a river can be destined for a hydropower operation downstream, where minimum flows are required to operate the facility. Currently, in a very few cases the amount of water in a river might be mandated by the state as minimum flows to protect the environment, including aquatic life in the river as well as downstream bays and estuaries.

Senate Bill 3 created a program to determine flow regimes necessary to support a sound ecological environment in each of the state's river basins and bays. A flow regime is a schedule of flows that reflects seasonal and yearly fluctuations for a particular basin. The TCEQ will consider recommended flow regimes as set-asides to be reserved for environmental flows when granting water rights permits. Some environmental organizations are concerned that water rights granted before these recommendations are put into law will not be subject to the set-asides developed by the process.

In some rivers, springs provide a substantial amount of the flow, and they can decrease or even dry up and thereby reduce the river's flow. The springflow reduction can be caused by drought and overpumping hundreds of miles away in an area that is unregulated. Because surface water in rivers and streams is disconnected legally and administratively from groundwater, surface water rights can continue to be used in most situations irrespective of whether contributing springs are flowing. This lack of conjunctive management in Texas presents us with one of the greatest risks to our water supply.

Lady Bird Lake is fed by the Colorado River. Formerly named Town Lake, it was renamed in 2007 in honor of Lady Bird Johnson and her efforts on the Town Lake Beautification Project. Photo courtesy of Texas Parks and Wildlife Department.

There is no guarantee that our rivers will always be there for us to use and enjoy.

DOES THE STATE HAVE TO LEAVE ANY WATER IN THE RIVERS?

Until as recently as May 2007, the answer to this question was essentially no. Texas has no overall policy for mandating that rivers flow to the sea. Since 1985 the state required that, to the extent practicable, new water permits within 200 river miles of the coast leave enough water flowing to maintain healthy bays and estuaries. The intent of this law was overshadowed by the fact that more than 90 percent of the water rights granted were pre-1985 and therefore had no protection for instream flow. In fact, some rivers already had more water rights granted than water in the river. Unfortunately, Senate Bill 3 did not include a catch-up provision to recover any of the water already permitted as set-aside for environmental flows.

It is useful to consider the various descriptive terms for environmental flows and what they mean. "Instream flow" is water in a river or stream necessary to support fish and wildlife and also refers to water needed for human recreational activities, such as canoeing, hunting, and fishing. Instream flows also protect water quality by providing dilution of contaminants. Sometimes the term "environmental flow" is used with essentially the same meaning. Once flow reaches the estuary it is referred to as "freshwater inflow"—flow from rivers necessary to preserve the health of bays and estuaries, which need dilution with freshwater. The reference to instream flows in this text implies that the flow should reach the estuary, but for the sake of brevity, the term "freshwater inflow" is not mentioned each time instream flow is mentioned.

IF YOU LIVE ON A RIVER DO YOU HAVE THE RIGHT TO USE WATER?

Due to laws based on residual English water law, persons in Texas who own land on the banks of rivers and streams have the right to a minimum amount of water—200 acre-feet per year—without a permit. This right, Domestic and Livestock Use, is only for landowner use for a home, farm animals, or gardens—not for commercial farms and ranches. This rule has been stretched to include amenities such as water skiing.

WHAT HAPPENS IN A DROUGHT IF THERE IS NOT ENOUGH WATER FOR EVERYONE?

Texas governs water under the doctrine of prior appropriation, or the "first in time is the first in right" rule. When a water right is issued it is given a priority date, and that date determines who gets the water first. Older, or senior, water rights must be satisfied before newer, or junior, rights. Junior rights are more likely to be terminated in a drought. In addition, the new environmental flow set-asides will be available for other beneficial uses during serious droughts.

CAN A WATER RIGHT BE BOUGHT AND SOLD?

A water right can be bought, sold, or leased. A water right does not have a title; in other words, it is not real property but rather a permit or license for use. A water right is granted in perpetuity, unless classified as a temporary permit. Obviously, more senior rights have higher value because they are the last to be curtailed in a drought. The physical location of a water right can also affect its value. The location cannot be changed without an amendment to the permit. A permit amendment to change the location of the diversion of the water requires that the TCEQ recalculate how the relocation will affect more senior water rights, and there cannot be a change in the availability of those existing rights.

WHAT HAPPENS IF YOU DON'T USE YOUR WATER RIGHT?

Officially, water rights can be canceled if they are not used for ten years, with some exceptions. However, cancellation is very rare; water rights are virtually treated as private property. Water rights stored with the Texas Water Bank are not subject to cancellation. Further, if the owner desires that the water be left in the stream for environmental flows, the right can be placed in the state's Water Trust.

DO YOU HAVE TO RETURN WATER TO THE STREAM?

Water legally diverted for a beneficial use does not have to be returned to the stream. The water can be used over and over. As the need for new sources of freshwater to meet demands grows, water reuse strategies are becoming more attractive to water users and planners. In fact, many power plants now use water repeatedly for cooling until it evaporates. Several Texas cities are reusing, or planning to reuse, large quantities of their treated wastewater to meet growing demands. This practice is intuitively sensible but can have negative effects on downstream interests.

The 2007 State Water Plan estimates that by 2060 reuse water will account for 14 percent of new water supply sources. Currently the laws regarding details of reuse permits are subject to much dispute. However the concept of reuse appears to be a permanent strategy in water planning. Further details of reuse strategies are presented in chapter 8 on the state water plan.

WHY CAN'T WATER BE MOVED FROM WET AREAS TO DRY AREAS TO SOLVE SHORTAGES?

Water can be moved from one river basin to another. This practice, interbasin transfer, has been used to meet water shortages for cities like Corpus Christi, which buys water from the Lavaca River. There are significant restrictions to this process that are hotly debated during each legislative session. Any water moved from one basin to another becomes junior to all the water rights in the system of origin. This means that during times of drought in the basin of origin, the transferred water right can be the first to be curtailed. Supporters of maintaining the restriction are often in rural areas that fear major cities will buy all of the water in their basin, harm agricultural production, and dry up their streams.

ARE THERE ANY RESTRICTIONS ON GROUNDWATER PUMPING?

Groundwater law is vastly different from surface water law in Texas, and in some areas it is practically nonexistent. Texans have the right to pump as much water as they can from beneath their property, unless the property falls within a groundwater conservation district. The right to pump water without restrictions, the rule of capture, does not require a permit, nor does it prevent one neighbor from overpumping and harming another's well. Texas is the only state in the United States without significant restrictions on groundwater use.

Groundwater conservation districts were first established by the Texas legislature in 1949. Though they vary in their rules, philosophies, policies, and regulations, they generally address issues such as the space between wells, permits for wells, and exporting water out of their areas. The first district was the High Plains Underground Water Conservation District, which governs a portion of the Ogallala Aquifer that has suffered decreasing levels due to pumping for irrigation. By the 1970s sinking of

The Civilian Conservation Corps completes "a job well done" in Big Bend's Chisos Mountains in 1934. Photo Courtesy of the Texas State Library and Archives Commission.

ground surface due to overpumping near Houston resulted in the creation of the first subsidence district in the nation. Until the 1990s there were not a significant number of districts, but since the passage of Senate Bill 2, numerous districts have been created.

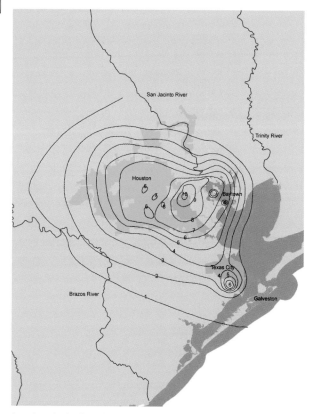

Approximate land surface subsidence in the Houston-Galveston Area due to groundwater withdrawals between 1906 and 1987. Lines indicate equal subsidence at one-foot intervals. Data Source: United States Geological Survey and Texas Natural Resources Information System.

One special groundwater district is the Edwards Aquifer Authority, which governs water in the Edwards (BFZ) Aquifer covering all or parts of eight Central Texas counties. The EAA was created in 1993 as the result of an endangered species lawsuit and mandated a cap on pumping from the aquifer to protect springflow in the San Marcos and Comal Springs. In 2007 the Texas legislature increased the pumping cap based on the historical permit pumping volume. At the same time the legislature is initiating a process with USFWS' Recovery Implementation Program to determine the necessary aquifer levels and spring discharge to preserve species. It is hoped that the result of this process will allow for the sustainable use of groundwater resources.

Groundwater districts do not cover the whole state, though many areas not covered are unpopulated or have little or no groundwater resources. About 90 percent of the groundwater pumped in 2000 came from areas governed by a groundwater district. By mid-2006 there were eighty-four confirmed districts and five awaiting confirmation. Of the 254 counties, 138 had all or part of their areas in a groundwater district.

Since the turn of the century, groundwater management has become more structured to facilitate statewide water planning. Since many groundwater districts were formed along political boundaries, such as county lines, their rules can conflict with adjacent districts that share the same aquifer, making long-term planning difficult. To address the varying rules and plans of groundwater districts, the legislature expanded the role of groundwater management areas (GMA). The state was divided into sixteen areas that generally follow aquifer boundaries, although in some cases other factors such as political boundaries were used. Some major aquifers were divided into multiple groundwater management areas based on hydrogeology and water use patterns. The groundwater districts in each management area have to agree on the desired future conditions of their area. These conditions include water levels of the aquifer, water quality, and springflows at a specified time in the future. The results are submitted to the Texas Water Development Board and, in most cases, are incorporated into regional water plans. Counties within a groundwater management area that do not have a groundwater district must still comply with the overall plan of the management area.

Groundwater districts have recently been challenged with projects seeking to buy or lease rural land and pump groundwa-

In 1991 Ron Pucek drilled the world's largest water well, capable of pumping 40,000 gallons per minute from the Edwards Aquifer for the Living Waters Artesian Spring catfish farm. By 2003 San Antonio had bought the farm and most of the water rights. Photo courtesy of Gregg Eckhardt and www.edwardsaquifer.net.

ter to municipal areas, sometimes hundreds of miles away. The city of El Paso has purchased several thousand acres about 150 miles from its city limits as a groundwater source. In 2005 the General Land Office planned to lease some of the state's public lands in West Texas to a private company to pump groundwater and transport it by pipeline to municipal users that were not identified. To date, such a deal has not been completed, but several entrepreneurial projects across the state envision just such a transaction.

Texas is gradually moving toward control of its groundwater. Starting with virtually no regulation in the early 1900s, there are now groundwater districts governing 90 percent of subsurface water resources. State water plans now include groundwater and surface water interaction in their overall predictions. Groundwater modeling is enabling groundwater conservation districts to more accurately predict the effects of future pumping and climate changes. Yet many challenges still exist for effective and sustainable management of groundwater in the future, including antiquated state laws and rules, the potential for vulnerable districts being controlled by parties interested only in export, areas that refuse to create groundwater districts, and the ever-present demand of large municipalities. A general lack of science and adequate resources compounds all these challenges.

WHAT ROLE DOES THE FEDERAL GOVERNMENT PLAY IN WATER RIGHTS?

Generally, the federal government stays out of water rights matters and leaves those decisions to the states. However, several groups have used the federal Endangered Species Act (ESA) to force groundwater regulation. The ESA was the basis of a lawsuit that led to the creation of the Edwards Aquifer Authority.

Water in West Texas makes a beautiful cut in the rock. Photo by Wyman Meinzer.

7. DOES TEXAS HAVE ENOUGH WATER?

Texas does not have enough developed freshwater supply to meet its future projected needs. This fact poses serious questions about the impact further development of freshwater will have on rivers, streams, springs, groundwater, and bays and estuaries. Furthermore, current systems, laws, rules, and science do not provide the tools necessary to effectively administer the water we do have. The 2007 State Water Plan estimates that if the state does not implement new water supply projects and management strategies there would be a water supply shortage of 8.8 million acre-feet by 2060. The plan recommends the construction of fourteen major reservoirs along with other projects, such as conservation, aquifer recharge, and desalination, to increase available water for future demands. While the plan proposes making up projected water shortages for human needs, it does not adequately address the issue of ensuring water for our rivers, bays, and estuaries. The state has sponsored numerous studies to determine freshwater inflow needs and collect instream flow data; however, these studies do not provide all the necessary information to make adequate management recommendations and are not fully accepted by the scientific community. Senate Bill 3 did create a process to establish and set aside flow regimes for the major bays and river basins.

INSTREAM FLOW LEGISLATION

The 2003 Texas legislature recognized the need for adequate science to determine instream flows and freshwater inflows and an improved administrative system to implement the science. In 2004 the legislature established the Study Commission on Environmental Flows with the statement:

> Maintaining the biological soundness of the state's rivers, lakes, bays, and estuaries is of great importance to the public's economic health and general well-being.

The Study Commission, in turn, appointed a Scientific Advisory Committee (SAC) consisting of expert engineers, attorneys, biologists, hydrologists, and economists to review the state of the science for environmental flows. Some of the main points in the SAC report presented in December 2004 are as follows:

- Concern that the present bay studies are not adequate and do not provide the flow information needed for the TCEQ to protect freshwater inflow in new water right applications.
- Different bays need different scientific approaches—not the current one-size-fits-all method.
- Criticism of the current statistical desktop methods for determining instream flow needs.
- Water is underpriced and does not reflect the value of environmental flows.
- Market-based solutions should be examined, such as incentives to place water in the Water Trust (a state-sponsored water conservation entity) or having the state acquire water rights to be placed in the Trust.

During the 79th Texas legislature in 2005, Senate Bill 3 was introduced that incorporated many of these recommendations. Considerable groundwork had been laid to gain consensus from all sides of the issue, including developers and environmentalists, but for various reasons the bill did not come to a vote. Subsequent special sessions in summer 2005 also failed to bring the water bill forward. The legislature appointed another environmental flows committee during the 2006 interim and considered recommendations similar to those that were proposed in 2005. During this time, TCEQ continued to manage water right applications using their historical case-by-case method as the only means of protection of flows in rivers and to bays and estuaries. In 2007 the legislature passed Senate Bill 3, incorporating most of the recommendations of the Scientific Advisory Committee and the Committee on Environmental Flows.

BAY AND ESTUARY AND INSTREAM FLOW STUDIES

In the 1980s scientists began creating complex models to determine the freshwater inflows needed to preserve the state's bays and estuaries. The state released the results of these studies in 1998, beginning with the Guadalupe Estuary and San Antonio Bay. Since then inflow needs have been calculated for all the major estuaries using the state methodology. However, the TCEQ has not directly applied the studies' results when considering new water rights and there have been many challenges to the validity of the studies themselves.

The Scientific Advisory Committee expressed concern that the bay and estuary studies were not adequate and did not provide the flow information that TCEQ needs when issuing water rights. Specifically, the studies mainly addressed the optimum flows needed for bays to remain productive areas for certain recreational and commercial fish, such as redfish, flounder, trout, and shrimp. The general choice of target species was defined by state legislation. Many independent scientists question this limited scope and would like to see a wider range of analysis to include even microscopic species, such as phytoplankton that form the basis of the food chain. In addition, scientists would

The Colorado River delta appears amorphous from the sky. Photo courtesy of Texas Parks and Wildlife Department.

like to see analysis of aquatic plant species that provide both a food source and habitat for many species. Species that dwell in the sediment of the bays are also important to the food chain and should be considered in the analysis of how much freshwater the bays need.

In addition to a broader perspective of the components of

the estuary protection, the SAC and many other scientists have asked for flow recommendations that are not just for optimum periods. What is needed are minimum flow levels for drought periods that would still enable the bay and estuary systems to survive. These types of flow recommendations would be helpful to the TCEQ in determining minimum flows in water permits.

Starting in 2003, Texas began a separate program analyzing the flows in rivers and streams. Trying to learn from previous shortcomings, the state is taking significant measures to ensure that the results of these studies are scientifically defensible and provide the answers needed to regulate water use while protecting the environment. The SAC was concerned that these studies, although thorough, would take so long that by the time the minimum flow standards could be determined, the state would have already appropriated those rights. Consequently, the SAC called for an intermediate system that could provide for an adaptive system to ensure flows while more extensive studies are completed. The instream flow studies will continue concurrently with the studies created by Senate Bill 3. Information gathered by the instream flow studies will likely be used in making the environmental set-aside recommendations.

THE CURRENT SYSTEM OF INSTREAM FLOW PROTECTION

When an application for a new water right comes before the TCEQ, the commission is required to include conditions needed to maintain the freshwater inflows necessary for bays and estuaries. Some of the limitations on this requirement are as follows:

- It has only been in force since 1985.
- It is only for permits within 200 miles of the coast.
- It is applied only to the extent practicable when all interests of the public are considered.
- There is no provision for environmental flows if there is not enough water left in the river from the previous granting of water rights.
- There is no definition in the law specifying the methods to

be used for establishing the necessary flows to protect the rivers and bays.

• There is no mandate to use the studies on environmental flows completed by the state, only that they shall be considered.

So what does TCEQ currently do about instream flow protection? Generally, to determine the quantities to be preserved, it uses a combination of the Lyons Method, the 7Q2, and in some cases the state's bay and estuary studies. These methods have been applied to varying degrees since 1985, and many groups, including the SAC, have criticized all three approaches.

The Lyons Method is a quick application, or desktop method, based on studies done on the Guadalupe River in Central Texas. It did not analyze flows needed for bays and estuaries, only instream flows for rivers. Despite its limitations, it is the most widely used method by TCEQ.

The 7Q2, which refers to the minimum seven-day, two-year discharge, is a statistically derived flow that forms the basis for water quality permits. A permit to discharge waste into a river has to satisfy the minimum standards of oxygen, temperature, and other chemical characteristics of water at the 7Q2 flow. According to the Texas Administrative Code, the 7Q2 was not intended for use as a minimum flow standard for preservation of biological species in water permits. However, it has been used for minimum flow requirements in some water right permits.

If most or all of the water in a section of river has already been granted, the TCEQ will sometimes deny a water right, and in fact produces a map showing the availability and lack of availability of water in certain areas. Even when only 75 percent of the water applied for is available less than 75 percent of the time—the period of historical records—the TCEQ can, and sometimes will, grant a new water right.

To examine the results of the TCEQ policy of granting water rights, studies have been conducted to show historical water rights granted and how many of these have environmental restrictions. Most of the water, or over 20 million acre-feet, was

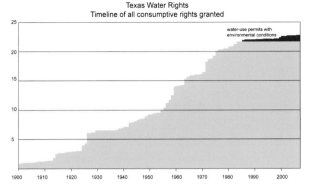

This figure illustrates the quantity in million acre-feet of consumptive rights granted in Texas over time. Data Source: National Wildlife Federation.

appropriated prior to the 1985 legislation requiring that flow be set aside for the environment. Only 8.4 million acre-feet of surface water is available in a record drought, according to the state water plan. Fortunately, at least for the time being, much of this historically granted water has not been utilized by the water rights holders, so many of the problems exist only on paper. Currently there are no plans to pass laws to revisit historical water rights granted before 1985 to apportion any for environmental flows.

THE WATER AVAILABILITY MODEL
Until the late 1990s, Texas did not have an adequate model that would determine how much flow would be left in a river under various scenarios, such as

- when all water rights are fully utilized;
- when no leftover water is returned to the stream;
- when only current water rights are used;
- when naturalized flows are considered as if there were no human impacts;
- when the impact of sediment gradually reduces reservoir capacity;
- when evaporation loss is considered;

- when the gain or loss of seepage into and from the river bank and bottom is considered; and
- when older or senior water rights having first priority in low flow situations are taken into consideration.

As of 2005, all of the major river basins in Texas have been modeled using the water availability model (WAM), which takes into account the science in the scenarios mentioned above. Simplistically, by building a theoretical model of what the naturalized flows with no human impact would be, a water right can be subtracted from the flow to determine what would be the remaining flow to reach the estuary. This model can be run assuming the climate conditions for any year of record, and in fact the TCEQ uses the record drought year for planning purposes. This model has been beneficial to the TCEQ for making determinations on the availability of water for a new water right. It has also been used to identify areas where there are true shortages and forecasting the future effects of unused water rights on instream flow and freshwater inflow.

One significant limitation of WAMs is that they do not address the groundwater and surface water interactions. Groundwater supplies the majority of base flows in a drought for some rivers, such as the San Marcos and Guadalupe. Up to 70 percent of the flow of the Guadalupe River at the coast during a drought of record is from the San Marcos Springs. However, the WAM does not provide accurate predictions of the springflows that enter the rivers. Groundwater availability models (GAM) are being perfected by the state that are expected to provide better information. Until these models are complete and integrated with the WAMs, there cannot be comprehensive predictions and planning for environmental flows.

HOW EFFECTIVE IS THE SYSTEM OF INSTREAM FLOW PROTECTION IN TEXAS?

Using information from the WAM, the TCEQ can compute the amount of flow that would remain in a river after all water rights are used to their full extent and no water is returned to the stream. The concept of return water is in itself controversial, and there is generally no mandate that water be returned to the stream after it is used. As water increases in value, some water experts anticipate that users will hold on to their water and reuse it over and over, or they will sell it after using it. For long-term water planning, these experts are in favor of assuming that no water will be returned to the stream. Other water experts favor assuming that 50 percent of the water will be returned. The TCEQ, when analyzing water availability for new water right applications, uses the assumption of no return flows.

One method of examining the level of protection of instream flow in Texas rivers is to calculate and analyze the flows that will reach the estuaries if all the water rights that have been issued were fully utilized. In many basins only a portion of water rights are currently being used. The National Wildlife Federation, in their 2004 report "Bays in Peril," performed such an analysis of the Texas bays and estuaries. A brief summary follows.

NWF first rated all Texas bays and estuaries by two measures:

1. The number of years that strong freshwater pulses were significantly reduced from what would have occurred in the natural state. These pulses are essential to the productivity of commercial and recreational species and generally occur in the first half of the year.
2. The number of six-month periods that fall below drought tolerance levels compared to the natural state. The drought tolerance levels were defined as being the minimum levels to sustain salinity for species survival as derived by the Texas Parks and Wildlife Department and Texas Water Development Board studies for the major Texas estuaries.

NWF then ranked the bays and estuaries as "Good," "Caution," or "Danger," based on the percent increase in at least one criterion of low flow resulting from all surface water rights being fully utilized. NWF assumed that 50 percent of the amount of

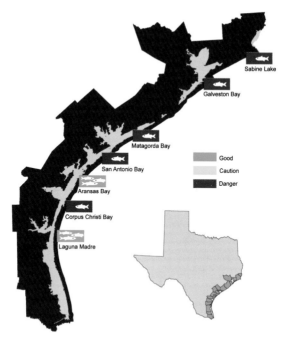

National Wildlife Federation classified bays in peril based on periods below drought tolerance levels and number of years with low freshwater inflow pulses. Data Source: www.texaswatermatters.org.

permitted water would be returned to the stream. A "Danger" rating means that the percentage of at least one of the low-flow criteria would increase by 67 percent or greater due to the full use of water rights. A "Caution" rating means an increase of at least one of the low-flow criteria from 33 to 67 percent. A "Good" rating shows a 0 to 33 percent increase.

Five of the seven major bays and estuaries received a "Danger" rating. Only Copano/Aransas Bay and the Upper Laguna Madre were rated as "Good." The Copano/Aransas Bay receives freshwater from the Mission and Aransas Rivers, which are relatively short rivers and have few significant water rights that would impact future flows. Total authorized water rights for the

Copano/Aransas system are only 1,900 acre-feet. Farther south, the Upper Laguna Madre is a hypersaline system with only a few small streams that provide limited freshwater. Due to its natural hypersaline state and the fact that there are less than 10,300 acre-feet of permits authorized on the streams that feed the estuary, the Upper Laguna Madre will not be seriously affected by reduced freshwater in the future.

NWF rated the Sabine Lake as an estuary in danger mainly because of a significant increase in the future of periods below drought tolerance levels. Water permits for the Sabine Lake watershed, including the Sabine and Neches Rivers, total about 4.6 million acre-feet per year, including 750,000 acre-feet from Louisiana. Currently, only about 1.2 million acre-feet of water are withdrawn per year from this watershed. A major threat to the freshwater supply of this basin is the potential transfer of water outside the drainage area.

Galveston Bay also received a "Danger" rating due primarily to increases in periods below drought tolerance levels. Existing water permits from the watershed, which includes the Trinity and San Jacinto Rivers, total 4.9 million acre-feet per year. Currently, only about 2.2 million acre-feet are withdrawn from the watershed. With the growing Dallas–Fort Worth and Houston metropolitan areas depending on this watershed, it will be a challenge to meet the freshwater needs of Galveston Bay in the future.

Matagorda Bay at the mouth of the Colorado River received a "Danger" classification. Matagorda Bay received one of the lowest ratings for freshwater inflow of any Texas bay, again mainly because of increases in periods below drought tolerance levels. The Colorado River, which provides a major portion of the freshwater for the bay, is under pressure from growing Austin in Central Texas and from rice farmers near the coast. There is also the proposed project to pump 150,000 acre-feet to San Antonio. Currently, about 1.36 million acre-feet per year of the 2.23 million acre-feet permitted are diverted. There are also additional applications for about a million additional acre-feet from the Matagorda Bay watershed.

The next bay to the south and west, San Antonio Bay, at the mouth of the Guadalupe and San Antonio Rivers, is another bay in danger due mainly to reduced periods of drought tolerance. Existing water permits total 651,000 acre-feet, and about 339,000 acre-feet are currently withdrawn annually. In spite of much fewer water permits and much less water withdrawal than the previously discussed bays to the east, San Antonio Bay still shows significant impact. The number of periods below drought tolerance in San Antonio Bay increase from two in the natural state to seven if all permits were fully utilized. In addition to surface water rights affecting freshwater inflow to the bay, the springflow from the Edwards (BFZ) Aquifer in Central Texas provides more than 70 percent of the baseflow to the Guadalupe and San Marcos Rivers in drought situations. In 2007 the legislature significantly increased the annual pumping cap, and there is concern that it will create more stress on the estuary than under prior rules.

Still farther west, in an even drier part of the state, lies Corpus Christi Bay, which receives flow from the Nueces River Basin. This bay is also ranked in the "Danger" category. The Nueces River is heavily affected by Choke Canyon Reservoir and Lake Corpus Christi, but there are mandates that limited amounts of freshwater have to be released to the estuary. These minimum flow requirements help to reduce the impact on periods that are below drought tolerance levels. However, the years with low freshwater pulses are still affected by the operation of the dams—thus the "Danger" ranking. Projects such as the pipeline from Lake Texana on the Navidad River and the Rincon Bayou freshwater inflow diversion are examples of how cities such as Corpus Christi have had to manage problems associated with a growing population in a dry area struggling with a limited water supply. The Lake Texana project brings in water for the city from a river basin to the east. The Rincon Bayou project redirects the already reduced freshwater flow at the mouth of the Nueces River to an area of the estuary that can benefit most from added flow.

With the passage of Senate Bill 3 and the establishment of a future system of determining and protecting flows to the bays, there is the possibility of preserving the ecological environment of Texas bays and estuaries. However, since the set-asides will not be implemented until 2010 at the earliest, there may not be enough unappropriated freshwater remaining to provide the necessary flows in a drought to support a sound ecological environment and maintain aquatic habitats in the estuaries.

THE WATER TRUST

In an effort to facilitate the preservation of instream flows, the Texas Water Development Board created the Texas Water Trust in 1997 to provide a place where owners of unused water rights can deposit their water rights and still retain ownership. The water right owner can specify a term for the deposit or leave it in the trust in perpetuity to protect instream flows. Although extremely rare in Texas, a water right can be canceled for nonuse; however, if the water right is deposited in the trust, this nonuse aspect does not apply. Another benefit to the owner is that the seniority date is preserved while the right remains in the trust.

By the end of 2006, only three water rights had been donated to the Water Trust. The first rights donated were for 1,236 acre-feet on the Rio Grande upstream of Big Bend National Park. In 2006 Texas State University at San Marcos made a donation of 33,108 acre-feet of San Marcos River rights to the trust. The university had purchased these rights as part of the acquisition of the San Marcos Springs headwaters in the early 1990s.

Although it is encouraging to see the Water Trust system in place and donations starting to come in, there are limitations in Texas water law that will probably prevent the trust from being a major force in environmental flow protection. First, it is voluntary, and there is currently no state funding for the purchase of environmental water rights. Second, the preservation of flow by a donated water right is only effective to the original point of diversion. For example, the donated water rights on the Rio Grande would not secure any water downstream of the point

The Rio Grande through Big Bend Ranch State Park was the recipient of the first water right donation in the Texas Water Trust. Photo courtesy of Texas Parks and Wildlife Department.

where the water was originally diverted. Downstream of the original diversion point, the donated water would be available for someone else to consume as a new water right. If there were no water left to be appropriated, as in most of the Rio Grande, then there would not be flow to protect most of the time. Still the donation on the Rio Grande prevents those water rights from being used and will increase river flow in some situations. The Water Trust is one of several ingredients of a complex solution to environmental flow preservation.

ENVIRONMENTAL FLOW APPLICATIONS

In 2000 the San Marcos River Foundation applied for a water right for instream flow for the Guadalupe River and the associated bays and estuaries as the sole intended use. Other environmental groups soon filed similar instream flow applications for Matagorda Bay, Galveston Bay, and Caddo Lake. Together these applications totaled over 6 million acre-feet. The TCEQ denied these applications, claiming that it did not have the authority to grant a water right solely for instream flow. The permit applicants filed lawsuits, claiming, among other things, that they had applied under the law and that TCEQ had no right to deny their applications. In 2006 the first court level ruled in favor of the applicants, saying they were entitled to an administrative hearing on the applications. Administrative hearings are part of the application process and do not necessarily mean that the permit will be granted. The TCEQ has appealed the lower court decision. These instream flow applications will probably be appealed all the way to the Texas Supreme Court.

PENDING WATER RIGHTS

As of 2007, there were over 4 million acre-feet of pending surface water right applications before the Texas Commission on Environmental Quality. Some of these applications were for wastewater reuse. This amount is equivalent to more than half of the surface water used in Texas in 2000, which was 6.83 million acre-feet. Without a clear direction from the Texas legislature on how to deal with instream flow protection in current permits, the TCEQ is struggling with what procedures it should use when granting the millions of acre-feet of pending water right applications from municipalities and river authorities. As a result, no new major water right applications have moved through the system for several years. The pending permits will be subject to existing TCEQ rules if authorized before the new Senate Bill 3 set-asides are created. Interestingly, there is a provision for increasing the amount of set-aside flows by 12.5 percent if there is new evidence showing that more flows are necessary. These newly approved permits might be subject to change to allow for increased environmental flow set-asides.

Texas is at a crossroads in its water supply, including water for the environment. The coming decades will most likely be a landmark era in Texas water policy.

Amistad Reservoir was created for flood control, irrigation, recreation, hydroelectric power, and water conservation. Photo by Wyman Meinzer.

8. PLANNING FOR THE FUTURE

In 1997, due to looming water shortages and an ongoing drought, the Texas legislature initiated preparation of the seventh state water plan since the 1950s. Unlike previous water plans, the current planning system is adaptive; that is, it is continually reviewed and updated in five-year cycles. The first plan was completed in 2002. In November 2006 the second plan, called Water for Texas 2007, was approved. The state plan is a compilation of the sixteen regional plans from the water planning groups and is administered by the Texas Water Development Board. The regional plans consider a moving fifty-year horizon. Thus the 2007 plan looks ahead to water needs in 2060.

The planning process begins with TWDB's projected population and water demand and engages consulting engineering firms to assist the planning groups. Each region determines how much water is available in the drought of record, how much is projected to be used in 2060, and what measures will be taken to make up the projected shortage, if there is one. However, the planning regions are not limited to proposing water manage-

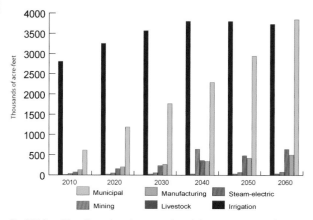

The 2007 State Water Plan estimated water supply needs by water use category for 2010–2060. Data Source: Texas Water Development Board.

ment strategies that approximate the projected need for their regions in 2060. For example, Region O in the Panhandle over the Ogallala Aquifer proposes a deficit of about 1.9 million acre-feet between its needs and the strategies for solving its needs. This deficit is mainly based on partial depletion of the Ogallala, which is very slow to recharge, and another source of water to replace the deficit is not proposed.

In other regions, the plans call for strategies that provide water in excess of their proposed needs in 2060. In Region C, containing Dallas, the plan calls for strategies that exceed needs by 730,370 acre-feet. Other regions with strategies exceeding needs are Region G on the middle Brazos River with 388,228 acre-feet; Region K, including Austin, with 304,619 acre-feet; and Region L, including San Antonio, with 315,940 acre-feet. Interestingly, when all the surpluses and deficits are added up they are almost equal. The total statewide needs in 2060 are 8.8 million acre-feet, and the total water strategies to meet those needs are 8.9 million acre-feet. Critics of the new water plan are concerned that regions can plan for more water than they propose needing in fifty years, including projects that may have detrimental effects on the environment, such as new reservoirs and excessive pumping of groundwater. The water plan describes these water management strategies as potentially feasible and does not appear to address the surplus or deficit issue.

Although still in their infancy, water availability models for surface water were available during the first five-year planning session. Improvements to the WAMs are ongoing. Groundwater availability models were not available for the 2002 planning process. Several models of major aquifers were completed during the 2007 planning process and were used to test strategy options. GAMs are being perfected for all major aquifers in Texas and will be incorporated in future state water plans. Another planning challenge in 2002 was the lack of complete studies determining the environmental flow needs for the bays and estuaries. These studies have since been completed but are the subject of many challenges regarding methodology, accuracy of data, and overall relevance of the results. Similar studies for environmental flow needs for rivers and streams are now under way. Senate Bill 3 flow regime study results will not be available for use in the planning process for several years.

SHORTAGES PREDICTED BY THE STATE WATER PLAN

The 2007 State Water Plan looks at predictions based on the drought of record, during the 1950s. During the planning process, areas of South and West Texas experienced a new record drought, which will be reflected in future plans. The prospect of climate change must now also be added to the equation. According to the 2007 water plan, by 2010 Texas will need an additional 3.7 million acre-feet of water. By 2060, due to the predicted increased population of 25 million, water shortage in a record drought is estimated to be 8.8 million acre-feet. Although the population is expected to more than double, the demand for water is anticipated to only increase by 27 percent, from 17 million acre-feet in 2000 to 22 million acre-feet by 2060.

The 8.8 million acre-feet of shortage does not include water for the environment—instream flows for rivers, springflows, and freshwater inflows for bays and estuaries. Sadly, and for various reasons, environmental flows are not among the uses included in the planning process. The regions that include coastal ecosystems performed varying degrees of analyses showing the impact

Region	Year and volume (acre-feet)					
	2010	2020	2030	2040	2050	2060
A	310,554	423,329	542,805	571,668	565,000	575,637
B	880	5,661	14,601	21,732	28,825	37,156
C	336,390	668,435	947,598	1,233,929	1,570,375	1,969,630
D	10,764	15,498	19,745	29,298	57,132	93,727
E	193,171	207,369	218,835	221,845	231,551	244,172
F	223,023	231,652	236,690	239,941	245,592	253,455
G	141,800	153,593	184,668	226,333	285,438	347,804
H	279,996	468,010	638,634	779,639	941,724	1,119,307
I	18,142	58,623	80,717	105,837	136,523	175,782
J	2,065	2,406	2,528	2,463	2,624	2,687
K	246,055	241,336	280,921	322,453	359,579	557,311
L	156,598	207,338	256,433	306,177	360,055	416,859
M	436,796	401,802	363,900	434,088	516,343	604,518
N	3,404	4,691	6,406	19,794	35,797	53,432
O	1,266,820	1,739,919	2,086,559	2,346,697	2,386,708	2,349,124
P	50,655	46,617	42,724	38,975	35,361	31,979
Total projected water needs	3,677,113	4,876,279	5,923,764	6,900,869	7,758,627	8,832,580

The 2007 State Water Plan estimated water supply needs for 2010–2060. Data Source: Texas Water Development Board.

of their projected strategies on freshwater inflows to bays and estuaries. The results varied from little or no impact to impact at lower flows or changes in the timing of flows. There was no consistent approach to the analysis by the coastal regions, and inconsistency among all sixteen regions with respect to the environment now appears to be a chronic failing.

The defined uses in the plan to calculate water needs are limited to the following:

- Municipal
- Manufacturing
- Mining
- Steam-electric
- Livestock
- Irrigation

Although projections show surface water availability in 2010 to be 13.3 million acre-feet, over 20.0 million acre-feet of surface water have been permitted by the state. Many of these water right holders will not be able to divert water in a record drought if their permits were issued more recently and have junior priority.

Groundwater Shortages

In 2003 groundwater provided 59 percent of the 15.6 million

acre-feet of water used in Texas. Seventy-nine percent of this groundwater was for irrigation. According to the 2007 State Water Plan, the estimated 2.7 million acre-feet of groundwater currently available with existing infrastructure will not be available in 2060 in a drought of record. This represents a 32 percent decrease in groundwater supplies.

This decline in groundwater supplies is primarily due to the depletion of the Ogallala Aquifer, which extends from northwest Texas to Nebraska and is very slow to be replenished. By 2060 the Texas portion of the Ogallala Aquifer is expected to decline by 2.5 million acre-feet. The Gulf Coast Aquifer is also expected to have reduced availability in 2060, even though availability has recently increased. Mandatory pumping limitations because of land subsidence will reduce the availability by 160,000 acre-feet. Based on these projections, planners can determine methods to solve the shortages, such as conservation, larger pumps, new wells, and pipelines.

Groundwater availability is another component of the water plan. Availability refers to how much water is available for use from an aquifer, not considering limits of existing infrastructure. Not all of the water in an aquifer is available according to laws, groundwater management rules, and planning group policies. The assessment of available groundwater by the regional planning groups is still a process in its infancy. Some regions used a policy of sustainability, or how much water can be pumped indefinitely without depleting an aquifer. Other regions plan for certain aquifers to be drained over a period of time. Region L is using a temporary value for the San Antonio segment of the Edwards Aquifer until a better value is developed through the Habitat Conservation Plan mandated by the U.S. Fish and Wildlife Service.

The previous state water plan estimated that only 10.1 million acre-feet of groundwater would be available in 2050, whereas the 2007 plan estimates 13.1 million acre-feet will be available in 2060. This difference is attributable mainly to the Far West and Plateau planning groups changing from a policy of aquifer depletion to one of sustainability. Also, in the El Paso area, the Hueco–Mesilla Bolsons Aquifer was originally forecasted to be depleted completely in the 2002 plan. A new groundwater model shows that, with current use, this aquifer is almost sustainable.

The changes in groundwater estimates are reflections of several aspects of groundwater planning. As new groundwater conservation districts are forming, more attention is being paid to sustainable pumping levels. Meanwhile, in some areas there are no groundwater districts and the rule of capture still reigns as groundwater is depleted. What ultimately happens in these areas may well be mandated by the legislature. Simultaneously with these policy developments, GAMs are being completed and refined, giving planners more accurate data.

There is still much information to be derived by groundwater models, including the relationship among aquifer levels, springs, and surface water flow. It is encouraging to see the sustainability concept being applied to various aquifers as part of the planning process.

Surface Water Strategies

By 2060 an additional 8.8 million acre-feet of water will be required in Texas. The 2007 State Water Plan proposes strategies that will provide 8.9 million acre-feet of additional water to make up the shortfall. The solutions to the total projected water shortage in 2060 are a mix of groundwater and surface water projects. Groundwater projects provide only 9 percent of new water supplies to solve the 2060 deficit. This leaves the burden of new water solutions on surface water resources, including the proposed fourteen new reservoirs, conservation, reuse, and desalination.

In the 2007 State Water Plan, surface water is proposed to provide 49 percent of new supply in 2060; the 2002 State Water Plan, in contrast, proposed that surface water would provide 66 percent of the new supply. This change is in part due to increased use of conservation, reuse, and desalination as water management strategies for 2060.

The largest category is "Other Surface Water," which provides 3,309,990 acre-feet of the approximate 9 million acre-feet of new water required and does not include water generated by new reservoirs. This other surface water category refers to various approaches to managing existing surface water, including

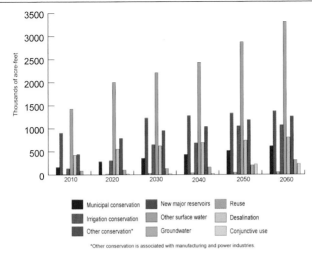

The 2007 State Water Plan estimated total new supply volumes generated by all recommended water management strategies for 2010–2060. Data Source: Texas Water Development Board.

new pipelines, purchasing additional water from major providers that have excess, obtaining additional water rights, reallocating water in existing reservoirs, and operating reservoirs in a given river as one system instead of as unrelated impoundments.

RESERVOIRS AND UNIQUE RIVER SEGMENTS

Surface water supplies available in a record drought are expected to decrease by 600,000 acre-feet by 2060, from 9 million to 8.4 million acre-feet. Of this reduction, 188,000 acre-feet will be caused by sedimentation in existing reservoirs, which reduces their capacity. The actual loss of reservoir capacity from sedimentation is much greater than this figure, but planning groups use only the firm yield, or amount of water available in a reservoir in a record drought. The firm yield greatly depends on the river or stream that feeds the reservoir and in some cases is only a rather small portion of the actual volume of a lake. Some reservoirs can fill up with a lot of sediment without much effect on firm yield. This disparity is shown by comparing the rate of sedimentation with the actual loss of firm yield in reservoirs. Estimates using new technology indicate that reservoirs in Texas

The Civilian Conservation Corps constructed numerous reservoirs around the nation during the 1930s. Photo courtesy of Texas Parks and Wildlife Department.

are filling up with about 90,000 acre-feet of sediment per year, or 0.27 percent of the volume. By 2060 this sediment will total 4.5 million acre-feet; however, the net result on firm yield of reservoirs is estimated to be a decrease of only 188,000 acre-feet.

The total volume of existing reservoirs that exceed 5,000 acre-feet in size is 34.47 million acre-feet. The current available firm yield is 8.9 million acre-feet. Although sedimentation appears not to be a major factor in future firm yield, there will be obvious reduced capacity effects on future reservoir management in nondrought years. In fact, this reduced supply to all reservoirs as a whole due to sedimentation exceeds the capacity of the fourteen new major reservoirs proposed by the 2007 State Water Plan. If built, the new reservoirs would provide 1.07 million acre-feet of new firm yield water available in a drought. So even though existing reservoirs are filling with sediment faster than proposed new ones would come online, the new reservoirs would provide an increase in water that would be available during a record drought.

Consideration has been given to dredging existing reservoirs, as opposed to constructing new ones. Currently, the cost of dredging reservoirs exceeds the cost of new reservoir construction. In spite of this cost variance, several smaller reservoirs have been dredged for reasons that include safety and navigation, as well as water supply.

In 1950 Texas had about 60 major reservoirs. Between 1950 and 1980, following the record drought, 119 new reservoirs were built, mainly for flood control; water supply was secondary. Since the 1970s reservoir construction has slowed, as reflected in the various water plans. The 1984 plan recommended 44 new reservoirs, the 1990 plan proposed 20, and the 1997 and 2002 plans each recommended eight. The reduction in reservoir construction is attributed in part to the fact that most optimum sites have already been used and there are substantial environmental problems with others.

The 2007 water plan proposes to designate remaining sites viable for reservoir construction as unique reservoir sites. These sites include the fourteen proposed reservoirs and additional sites recommended by regional planning groups. The Texas Water Code provides for this designation, which prevents a state agency or political subdivision of the state from buying the site or obtaining an easement. This designation does not provide total protection for a future reservoir site; for example, some federal designations may supersede the state designation. In North Texas, Region C (including Dallas–Fort Worth) proposed the Fastrill Reservoir in the 2007 plan. Some of the land that the reservoir will inundate was designated part of the Neches River National Wildlife Refuge in summer 2006. National wildlife refuges are administered by the federal government. If this designation stands up to possible challenges, the site would be unavailable for the proposed reservoir. In early 2007 the Texas Water Development Board and the city of Dallas filed suit against the U.S. Fish and Wildlife Service over the designation of this preserve and its potential conflict with the Fastrill Reservoir.

There is also a special designation for rivers and streams of unique ecological value in the Texas water plans. The basic principle of this provision is protection from reservoir construction or other adverse development on the designated segments. In addition, all new reservoir projects are required to provide protection of an equivalent or greater area, called mitigation, to compensate for the wildlife habitat destroyed by the project. The new water plan requests that the legislature provide assistance in funding the acquisition of new reservoir sites and any mitiga-

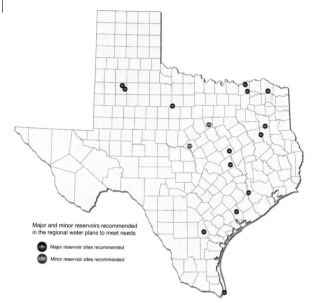

Location of the recommended major and minor reservoir sites in the 2007 State Water Plan. Data Source: Texas Water Development Board.

tion areas. The plan also requests that the legislature designate those river and stream segments of unique ecological value that are recommended in the 2007 plan as mitigation for future reservoirs. Only two planning groups recommended river and stream segments of unique ecological value, Region H (Houston) and Region E (Far West Texas), an indication of the level of concern for the environment among the regional planning groups.

WATER CONSERVATION

Consideration of water conservation as a source for future supplies received more attention in the 2007 State Water Plan than in the previous five-year plan. Total water supply from conservation is proposed to be almost 2 million acre-feet, or about 23 percent of the total new supply required in 2060. The 2002 plan proposed only 990,000 acre-feet, which was 14 percent of the water needed in 2050. Active conservation measures are actions initiated by utilities, businesses, water customers, and agricultural interests to reduce water consumption. Passive conservation is already built into TWDB's planning numbers. Passive water conservation refers to water saved by state and federal regulation that requires water-efficient plumbing fixtures, including showerheads, faucets, and toilets. Over time, more of these types of fixtures will be installed as old fixtures wear out and are replaced by new efficient ones. The TWDB estimates that by 2060 passive conservation will reduce municipal demand by 6.6 percent, or 587,000 acre-feet. This will lower per capita per day consumption by 11.5 gallons statewide.

In 2003 the state legislature laid the groundwork for conservation by creating the Water Conservation Implementation Task Force. The task force undertook extensive analyses and made recommendations for conservation that included target consumption per person, or gallons per capita daily (gpcd). It also recommended a number of strategies and methods for conservation—Best Management Practices (BMPs)—for municipal, irrigation, and industrial water users.

Ecologically Significant Stream Segments

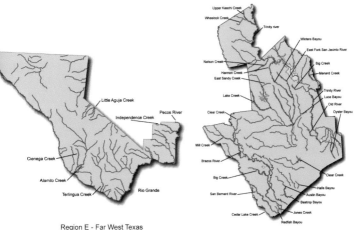

Region E - Far West Texas

Region H

Region H and Region E were the two planning groups to designate river and stream segments of unique ecological value. Data Source: Texas Parks and Wildlife Department.

MUNICIPAL CONSERVATION

Municipal conservation entails social and technical approaches. Examples of social approaches are water pricing structures that impose higher rates as use increases and educational programs that encourage conservation. Technical approaches include switching to more efficient plumbing fixtures and replacing aging water distribution systems, including leaking pipelines. Many cities offer rebates for trading in older, inefficient fixtures for newer efficient ones. The plan summarizes supplies generated by municipal conservation as to their amount, percent of all supplies, and cost. The cost per acre-foot for municipal conservation is relatively inexpensive, averaging $234 per acre-foot. This can be compared to the average cost for new reservoirs, $374 per acre-foot.

Many of the conservation strategies in the 2007 State Water Plan were based on recommendations of the Water Conservation Implementation Task Force. One of these was a target of 140 gpcd for municipal users. If a user group was using more than the 140 gpcd, the task force recommended reducing their usage by 1 percent per year until it reached 140 gpcd. Some of the regional planning groups use this 140 gpcd as a target.

A controversial aspect of the task force's recommendations involved the interpretation of wastewater reuse. In spite of objections from the Sierra Club and some task force members, the final report recommends that wastewater reuse be credited to a municipality as conserved water. The dissenters contend that water reuse is not actual conserved water. The argument is that if a person continues to use the same gpcd after some of it is reused, the same amount per person is still being consumed, and that is not true conservation.

DIFFERENCES IN ESTIMATED USAGE PER PERSON Another aspect of the water plan that arouses concern involves the varying amounts of per capita use on which cities can base their future plans. To demonstrate this point, the National Wildlife Federation has calculated the projected gallons per person per day for the ten largest Texas cities in 2060. These numbers do not include credit for reused water, only water that is conserved through other measures. Dallas shows the highest per capita usage, followed by Fort Worth. The projected amounts for 2060 vary from 233 gpcd in Dallas to 124 gpcd in San Antonio. Based on these numbers and the estimated population increases, the planning groups propose various water management strategies to satisfy projected unmet needs.

The water plan offers an explanation of this variability in per capita water use, citing the following reasons:

- Climate variations, including rainfall and temperature
- Specific year weather patterns for the base year 2000
- Commercial and institutional water use
- Type of residential development
- Income of customers
- Seasonal residents
- Age of infrastructure

The water plan also notes that the method of calculation of per capita use can vary from one city to the next; therefore, comparison of one city's usage to the next should be approached with caution.

The National Wildlife Federation has compared the need for proposed water infrastructure projects with possible increased water conservation. One of the projects examined was the proposed Brownsville Weir, a dam on the Lower Rio Grande. This dam, proposed by Region M, which includes the Lower Rio Grande, is expected to provide an additional 20,643 acre-feet of firm yield water. The Water Conservation Implementation Task Force recommended that Brownsville conserve water by reducing its annual usage by 1 percent. NWF claims that if the city performed the recommended conservation, starting with 2005 usage levels, they would achieve the target 140 gpcd in 2060. The result would be an additional savings of 28,852 acre-feet in 2060, exceeding the production of the proposed reservoir. It is worthy to note here that for a period of time in 2001 the Rio Grande stopped flowing to the Gulf of Mexico, arousing concern about the feasibility of an additional dam on the river.

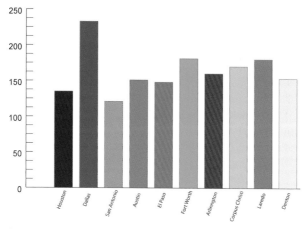

Daily per capita water usage in ten Texas cities. Data Source: National Wildlife Federation.

It is encouraging to see the increased emphasis on municipal water conservation in the 2007 plan as compared to the 2002 plan. Moreover, with further conservation many expensive projects could be avoided, not only saving money, but also preserving rivers, streams, wetlands, bays, estuaries, and other natural areas of Texas.

Agricultural Conservation

Currently, irrigation comprises 60 percent of all water use in Texas. However, by 2060 irrigation is expected to decline to 40 percent. Twelve planning regions call for future irrigation conservation to help meet demand. Some of the recommended strategies include:

- managing irrigation water use through scheduling, measurement, conservation tillage, and crop residue management;
- managing the land with furrow dikes, leveling, conversion to dry land farming, and brush control;
- improving farm delivery by lining ditches and installing low-pressure center pivot sprinklers and drip systems;

- improving water district systems by lining canals and replacing canals with pipes; and
- recovering and reusing irrigation water.

All irrigation conservation strategies are expected to provide 1.4 million acre-feet of new water in 2060, which is about 37 percent of total irrigation needs. Regions A, O, and M produce 80 percent of irrigated crops in Texas. In 2060 irrigation conservation in these three regions will produce about one million acre-feet of water.

Costs of irrigation conservation are relatively low, from $1 per acre-foot to $216; however, many of these costs would have to be absorbed by the farmers themselves—not spread out over a larger entity, such as a water district. This cost burden could either prevent or delay implementation of irrigation water saving measures. The TWDB has several projects to demonstrate various conservation strategies, and many planning groups have recommended additional state and federal funding for agricultural conservation programs.

Agriculture conservation appears to be a significant source of water in Texas' future, ranking third in total water developed in 2060 compared to all other proposed sources.

GROUNDWATER MANAGEMENT STRATEGIES

Groundwater management strategies are estimated to provide about 800,000 acre-feet, or about 9 percent, of new water supplies in 2060. Currently, there is more emphasis on managing aquifers to a sustainable level than depleting them. Still, there are some areas where drawdown of certain aquifers such as the Ogallala is part of the long-term plan. Other regions propose temporary overdrafting during drought in their plans. Other methods to increase groundwater availability are new wells, increased pumping from existing wells, and transfer of groundwater from areas of surplus to areas of need. One of the most hopeful strategies for protecting and enhancing groundwater supplies is the aggressive effort by San Antonio to protect the recharge area of the Edwards Aquifer.

REUSE STRATEGIES

Water reuse is receiving a great deal of attention as resources grow more scarce. In the 2007 water plan, reuse accounts for 1.3 million acre-feet in 2060, which amounts to 14 percent of new water supply. Previously, the 2002 plan proposed 420,000 acre-feet, amounting to 6 percent of new water supply in 2050.

Reuse can be direct, which means piping the wastewater directly from the treatment plant to the point of use, or indirect, whereby treated effluent is discharged into a river and diverted downstream for use. Currently, the laws regarding details of reuse permits, especially indirect reuse, are subject to much dispute, resulting in a backlog of pending permits at the Texas Commission on Environmental Quality. However, the concept of reuse appears to be a permanent strategy in water planning.

One of the major advantages of reuse is that it is drought-proof. As long as a city has wastewater, there is water to be reused. Also, reuse is the only water source that automatically grows as population grows. Unlike most water sources, reuse is normally located near the point of use and therefore does not have to be transported long distances. Region C has the largest amount of projected reuse, 722,320 acre-feet, which is about 57 percent of proposed statewide reuse projects in 2060. The average cost per acre-foot for all reuse strategies is estimated at $248, making it one of the more economically competitive management strategies.

Reuse as a strategy for future water supply has many possibilities, but like most strategies, it has potential problems. Two of the biggest concerns are the possible effects on the environment and on other downstream water rights holders and uses if previously discharged water is no longer discharged. These problems will have to be worked out before reuse can become a major water supply strategy. With numerous pending reuse applications before the TCEQ, the solution to these problems could either come soon or end up in lengthy court deliberations. The 2007 State Water Plan also recommends that the legislature develop a policy to clarify issues surrounding indirect reuse.

DESALINATION

Desalination, or the removal of salt from brackish groundwater or seawater, was included in eight regional water plans. The projects proposed would create 313,000 acre-feet of water, with 44 percent from seawater and 56 percent from brackish groundwater. This total is up from 180,000 acre-feet proposed in the 2002 plan, which results from an increasing awareness of desalination combined with reduced costs as the technology advances. The cost per acre-foot for seawater desalination ranges from $768 to $1,390; brackish water desalination costs range from $429 to $953. In most cases the brackish water has less salt concentrate than seawater.

Although one of the first significant desalination plants in America was established in Freeport in the early 1960s, when John F. Kennedy was president, currently in Texas there are no seawater desalination plants. There are, however, thirty-eight desalination plants in Texas with a capacity greater than 25,000 gallons per day that process brackish groundwater and have a combined capacity of 52 million gallons per day. Two pilot plant studies are being conducted in the Lower Rio Grande Valley that could result in a full-scale saltwater desalination plant in the future. There are approximately 250 desalination plants in the United States. Worldwide, there are 12,500 desalination plants in 120 countries with a capacity of 4 billion gallons per day. Sixty percent of these are located in the oil-rich Middle East, where the cost of water is a secondary issue.

The environmental impact of desalination projects will have to be addressed before they can be considered a viable source of future water in Texas. On first impression, the supply of seawater is considered almost infinite and a relatively inexpensive source of water. However, there are concerns about the disposal of the concentrate resulting from the process. In a reverse osmosis plant the concentrate can be as much as two times as salty as seawater. Options for disposal include discharge into the bays, piping out to the sea, deep well injection, and land application. The concentrate, being heavier than seawater, tends to sink when discharged and can cause low oxygen levels that

Reverse osmosis equipment for seawater desalination at the pilot desalination plant in Brownsville. Photo by Dr. Hari Krishna and courtesy of the Texas Water Development Board.

harm aquatic life. This is especially true for bays, which are generally confined areas with less circulation than the open sea. Unfortunately, disposal of concentrate into the bay is often the most economical method. To lessen the effects of direct discharge into a bay, some desalination plants discharge through extended pipelines that reach out into the open sea. Inland disposal has the potential to contaminate groundwater. Extensive studies of the areas that may be affected are necessary before these methods can be utilized.

It is likely that we will have to rely more heavily on water produced by the desalination process as a future source of water. As Texas rivers become overappropriated, effective reservoir sites are built out, groundwater is pumped below sustainable levels, and conservation achieves maximum acceptable limits, the sea is one of the last sources of water. But desalination is not a benign process. Several potentially negative side effects can result from discharging highly concentrated salt water back into the environment.

CONJUNCTIVE USE

Another water management strategy belatedly receiving a great deal of attention is conjunctive use, or the combining of surface water and groundwater to optimize both sources. Often this strategy calls for surface water to be the primary source and for groundwater to be used only during a drought. Region L proposed the largest conjunctive use: 180,000 acre-feet by 2060. Region L projects make up the majority of the 231,000 acre-feet of proposed conjunctive use projects in the 2007 State Water Plan. The project proposed by LCRA calls for 150,000 acre-feet of Lower Colorado River water near the coast to be conveyed by pipeline to San Antonio. This project, already in the early planning stages, is meeting some resistance. Many opponents are concerned about the use of groundwater for agriculture during a drought, as well as decreased flows to Matagorda Bay and the impact of the off-channel reservoir.

Conjunctive use is a strategy that is here to stay, but it requires extensive analysis of surface and groundwater resources and the effect of a drought on their firm yield. Too often the interrelationship of ground- and surface water has been ignored. In conjunctive use projects, where the sources of ground- and surface water are connected hydrologically, what affects one source often affects the other. For effective and sustainable conjunctive use projects, detailed modeling will have to be performed to determine the long-term effects of using interrelated sources as supply during record droughts. If the sources are not connected hydrologically, water planners need to look at the effects on the aquifer of origin or surface water feature of origin due to increased consumption from either.

LAND STEWARDSHIP

One of the most talked about potential sources of new water in Texas is brush control, which involves removal of vegetation that uses large amounts of water. Brush control is one of many land use strategies that improve the quantity and quality of surface water and groundwater. Other strategies are maintaining riparian or streamside vegetation, reseeding native plants, maintaining open space and wildlife habitat, conserving wetlands, and controlling erosion by preventing overgrazing.

Brush control in Texas usually involves the removal of Ashe

juniper or cedar, mesquite, and salt cedar or tamarisk. These types of brush and scrub trees use large amounts of water through evapotranspiration, or the wicking of water into the atmosphere through their leaves. In addition, Ashe junipers catch and hold a lot of rain in their canopies, which are green year-round, causing more water to be lost to the atmosphere. This results in little or no vegetative cover between these types of trees. A good grass cover will slow down runoff and enable it to be stored in the soil and recharge the groundwater, as well as provide baseflow (the portion of streamflow that comes from groundwater) for streams and rivers. A lack of ground cover will increase runoff. Salt cedar, which grows along the edge of streams, affects streams in a different manner. It not only uses water in the soil but also consumes water directly from the stream. Studies have shown that if salt cedar is removed from stream banks, streamflow increases.

Although studies on the effect of brush control are ongoing, there is still no agreement on the quantity of water saved by removal. Therefore, the regional water plans do not include any future new water supply based on these strategies. However, Region G, in north-central Texas, including parts of the Upper and Middle Brazos River, recommended brush control as a management strategy, yet the 2007 water plan did not include any supplies based on this strategy. We hope that this lack of official recognition in the state water plan will not discourage the implementation of effective land use strategies that improve the quantity and quality of water in streams while also preserving terrestrial and aquatic habitat. Many private landowners in Texas are already incorporating these methods into farming and ranching operations, and this should be encouraged by the state.

STREAMFLOW ASSESSMENT

In response to numerous requests, the 2007 State Water Plan includes estimates of the impact of proposed water management strategies on streamflow. The TWDB selected 156 sites as the control points for analysis, and by using the WAM, they compared current streamflows to 2060 monthly median and tenth percentile flows. Tenth percentile flows are computed by sorting

Salt cedar can be seen in black after brush control practices along the Pecos River. Photo courtesy of Texas A&M University.

all monthly flow values to find the monthly value for which only 10 percent of recorded flow values are less, as would occur in serious drought. Median monthly flow values are computed by finding the monthly value for which 50 percent of recorded flow values are less, as would occur under more normal conditions.

Future flow assessments can show decreases from natural causes such as drought, as well as changes in flow due to return flows from new water projects, new water supplies being moved down a river from one area to another, changes in points of diversion, and many other factors. Also, seasonal flows can often change due to upstream reservoir management, resulting in storage of water in wetter months and release of water for downstream users in drier months.

As long as new water management strategies proposed in a regional water plan are analyzed using what is called the consensus criteria, the impact on streamflow is acceptable. If a stream is already heavily affected by water diversions implemented before the planning process, there are really no criteria in place. The criteria apply only to new projects proposed in the plans. These criteria were established by state environmental agencies, various experts, and citizens to serve as a desktop approach estimating environmental flow needs when planning new reservoirs or diversions. Also, when the TCEQ considers a new water right it is not required to use the consensus criteria since they were developed for planning rather than permitting.

SB 3 requires a more comprehensive process to achieve a consensus-based, regional approach to integrate environmental flow protection with flows for human needs.

GLOBAL WARMING AND THE STATE WATER PLAN

While the Texas water plan provides a forum for envisioning water management fifty years from now, some critics are concerned that the vision is not broad enough because it does not project the effects of global warming. Water needs in fifty years are generally based on climate constants and the assumption that the 1950s drought of record is the worst-case scenario. The reality is that history shows droughts of much greater duration and climatologists forecast future droughts of increasing severity. Most scientists agree that global warming is already affecting our climate patterns and sea levels, and there will be noticeable effects on our water supply in the foreseeable future. Global warming is even being recognized by the insurance industry in the form of policies being canceled in high-storm-risk areas. Some major insurance companies are encouraging reduction of greenhouse gases when rebuilding after disasters, recognizing that it is good business in the long run to reduce the sources of global warming. Even the Pentagon, in its 2004 study on security issues related to global warming, recognized that global warming could seriously affect the planet.

With this widespread agreement that global warming is a problem, many are calling for various aspects of global warming, including reduced rainfall and runoff, to be included in the state water plan. Gerald North, professor of geosciences at Texas A&M University, participated in studies with the National

King Ranch's drought-prone rangelands have required innovative water management strategies. Photo courtesy of Texas Parks and Wildlife Department.

Academy of Sciences in 2006 that found the last few decades of the twentieth century were warmer than any period in the past four hundred years. North reported that computer estimates show a 4 to 9 degree F increase in temperature by the end of the twenty-first century. In addition, a simulation by the University of Texas showed that a 4 degree temperature rise and a 5 percent decrease in rainfall would result in a 25 percent reduction in runoff in Texas. The same model showed river flow decreasing by one-third, with an even larger decrease in a drought.

North has expressed serious concern that the state water plan does not include global warming factors in its estimates of future water availability. The Texas Water Development Board says that the agency reassesses water needs every five years and that adjustments are made for climate change. Perhaps with the growing awareness of global warming, the participants in the next planning cycle will look further into the future and consider the looming presence of global climate change on our water resources.

The Lower Trinity River provides many beautiful swamp habitats. Photo by Wyman Meinzer.

9. WHAT'S IN YOUR WATER?

IS TEXAS WATER BEING MONITORED?

Keeping track of water quality in a state as large as Texas is a daunting task. Texas has almost 200,000 miles of rivers and streams (of which about 40,000 miles have year-round flow), 6.5 million acres of inland wetlands, 1.7 million acres of coastal wetlands, 3 million acres of reservoirs and lakes, 2,394 square miles of bays and estuaries, and 3,879 square miles of open Gulf under its jurisdiction. Regular testing and monitoring of all these water bodies is impossible, so the state must make choices about which areas to test and how often. With assistance from river authorities, the state conducts annual stakeholder meetings to discuss river segments that might have pollution problems, and then state agencies select segments to assess that year. In 2004 Texas assessed about 11 percent of its total river miles, 80 percent of lake area, 85 percent of bay area, and 100 percent of open Gulf jurisdiction. The state also monitors groundwater, but with more than 800,000 water wells drilled in the twentieth century and 20,000 new wells per year, there are simply not enough resources to test them all.

The state or federal government establishes, monitors, and enforces water quality standards for these assessed water bodies. There are three main categories of water quality standards:

1. Stream standards or surface water quality standards—established by the state and used to determine the amount of various substances that can be discharged into and assimilated by streams. The standards must comply with federal minimum requirements and be approved by the U.S. EPA.
2. Effluent standards for wastewater discharges—set by the state and administered through wastewater permits.
3. Drinking water standards—a combination of state and federal guidelines for drinking water supply after its withdrawal from a stream or aquifer.

WHAT TYPES OF POLLUTANTS ARE IN THE WATER?

If one considers chemicals individually, there are thousands of different pollutants, and significant synergistic effects from their many combinations, that can enter our water supply. Each year more than a thousand new chemicals are developed and introduced into the environment. Estimates show that there are currently about one hundred thousand chemicals on the market, and it is impossible to monitor all drinking water sources and each waterway for all of the potential chemicals and other pollutants. Testing for some of these new chemicals is very expensive and is not performed on a regular basis. Only a few hundred chemicals are monitored in public water supplies. Far fewer are monitored for segments of rivers and reservoirs not directly associated with drinking water supply. Municipal water supplies are responsible for annual reports on numerous pollutants in their water systems, but they cannot keep up with the growing list of chemicals that we create and discharge.

In addition to chemicals, there are harmful pathogens, bacteria, and toxic algae present in some water bodies. Following is a brief discussion of the types of pollutants, where they occur, some of the problems they cause, and the measures taken to reduce the contamination.

Pathogens—A general term referring to any organism that can cause disease. This includes bacteria, fungi, viruses, and parasites.

Fecal Coliform—This is a bacteria group, the presence and concentration of which is often the basis for determining whether swimming areas and drinking water are safe or not. Fecal coliform bacteria live in the digestive tract of warm-blooded animals—humans, pets, farm animals, and wildlife. These bacteria are generally not harmful to humans, but they indicate the presence of other disease-causing bacteria, including those that cause typhoid, dysentery, hepatitis A, and cholera. Direct testing for these harmful pathogens is expensive and impractical because the pathogens occur sporadically and usually at low levels. It is more efficient and economical to test for the fecal coliform group of bacteria. Some states are now performing tests for a specific member of the fecal coliform group, *Escherichia coli* or *E. coli*. *E. coli* testing is time-consuming and expensive; however, EPA studies show that *E. coli* is a good indicator of harmful bacteria. The state of Texas bases its analysis of water on either *E. coli* or fecal coliform standards. In order for the state to classify rivers and lakes as contact recreation areas (swimming), the levels of fecal coliform and *E. coli* must not exceed the maximum standards. There are numerous stream and river segments in Texas that do not meet the EPA standards for fecal bacteria.

Elevated levels of bacteria are one of the major causes of pollution in surface water. Shellfish, such as oysters in some bays, have shown elevated bacteria levels, causing the Texas Department of Health (TDOH) to issue warnings. In 2002 the TDOH found that fifteen of the thirty-two bay segments designated as oyster waters had elevated fecal coliform contamination levels.

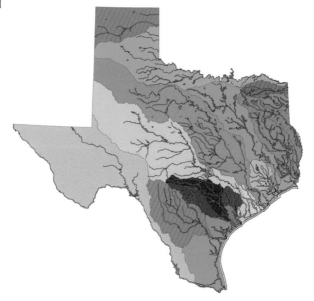

The Texas Commission on Environmental Quality listed these stream segments as the Impaired Waters of Texas. Data Source: Texas Commission on Environmental Quality.

These areas included all of Nueces Bay, Cedar Lakes and Chocolate Bay, and major portions of Upper Galveston Bay, Trinity Bay, Lower Galveston Bay, and Carancahua Bay.

In the Bosque River watershed there have been problems with intensive cattle operations, or Concentrated Animal Feeding Operations (CAFO), whose animal waste runoff pollutes the river and results in high levels of fecal coliform. Programs are under way to reduce this pollution source by requiring treatment of the runoff from these operations.

The good news is that fecal coliform is intensely monitored in drinking water and that chlorine, ultraviolet light, and ozone in sufficient quantities will kill these bacteria.

Cryptosporidium—This is a parasite that can cause death and unfortunately is resistant to chlorine. In 1993 an outbreak in

Milwaukee affected 400,000 residents and caused at least 50 deaths. Cryptosporidium is strictly monitored in water supplies, and since its discovery there have been few outbreaks. If found the parasite can be eliminated with a very fine filter or killed by boiling water.

Nutrients—It is interesting that the two main chemicals used as basic fertilizer for the crops we grow, nitrogen and phosphorus, are considered pollutants when introduced in excessive amounts to rivers, reservoirs, and groundwater. The same nutrients that benefit agricultural crops can cause excessive algae to grow in the stream, which results in a lower amount of the oxygen that is needed by fish and other aquatic species. As algae grows, it reduces the amount of light that can pass through the water, thereby affecting plant and animal life. Nitrates, a compound of nitrogen and oxygen, in quantities above drinking water standards are also harmful to human infants and can cause the blue baby syndrome that can be fatal.

Sources of nutrients can be excess fertilizers, CAFOs, wastewater plant discharges, and runoff containing animal feces. Most wastewater plants in Texas contain facilities to remove much of the nitrogen. Phosphorus is only recently being addressed as a problem nutrient in discharges, and it is often only through citizen support that wastewater treatment plants remove it.

In the Bosque River Basin there are an estimated 180 CAFOs with over 130,000 head of cattle producing 1.8 million tons annually of manure containing nitrogen and phosphorus. In order to prevent excessive nutrient levels, most feedlots are now required to have permits and must not allow the runoff to enter the waterway. There is so much manure in the Bosque watershed, which provides drinking water for the city of Waco, that the state is managing the Composted Manure Incentive Project. This project provides transport and other incentives to move manure out of the basin to be composted. Between 2000 and 2003 over 685,000 tons of manure were removed under this project.

TCEQ requires that some confined animal feeding operations (CAFOs) obtain water quality permits. Photo courtesy of the United States Geological Survey.

Metals—Many metals, such as calcium, magnesium, potassium, and sodium, are essential to sustain human life. In much smaller amounts, we also need cobalt, copper, iron, manganese, molybdenum, selenium, and zinc. However, high amounts of these can be hazardous to our health. Many metals are toxic and need to be kept out of drinking water entirely, including aluminum, arsenic, barium, cadmium, chromium, lead, mercury, and silver.

Among the sources of metals are discharges from industry and household wastes dumped into wastewater systems. Lead is a major concern, especially because of its effects on the development of children. Sources of lead include old lead pipes in homes, brass fixtures, or solder in copper pipes. The use of lead solders in plumbing fixtures is now banned by the EPA, but some old lead pipes and lead solder still exist in older homes.

Mercury in water is a widespread problem in Texas and in several other states. A primary source of mercury is airborne contamination, or acid rain, from coal-burning power plants. Measures are under way to reduce coal plant emissions, especially from lower-grade coal, which contains more mercury than higher-grade coal.

Uranium contamination from mining is a problem in the groundwater supplies of Live Oak and Karnes Counties where a former strip mine was operated. Groundwater continues to show traces of uranium and is being monitored. In Goliad County a private firm drilled test wells in 2006 to analyze the potential for a new uranium mine. There is much local opposition due to concern about groundwater quality, including a resolution by the county commissioners opposing any on-site uranium mines in the county.

In various parts of Texas, barium, arsenic, and cadmium are a by-product of oil production. Since 1969 regulations have been in place requiring permits to dispose of these materials. The Texas Railroad Commission, which oversees oil production, reported 225 cases of groundwater contamination from oil and gas in 81 counties in 2002. In addition, the Oil Field Cleanup Fund identified 1,629 abandoned oil field sites in 2002 that are candidates for state-managed remediation.

Conventional wastewater treatment systems are not designed to deal with metals or other toxic chemicals. Most cities require that industrial facilities pretreat their wastewater to limit these toxins to acceptable levels before their discharge enters the wastewater system. Larger cities are required to maintain and enforce a pretreatment program, but there is no guarantee that toxins do not enter the rivers and streams.

Toxins/Carcinogens/Organics—These are chemicals that can cause death (toxins) and cancer (carcinogens), some of which are man-made and consist mainly of carbon (organics). Organics include many pesticides and can be toxic or carcinogenic. There are so many of these substances that it is impossible to test for each one of them. One method used by the Texas Commission on Environmental Quality, biomonitoring, uses particular aquatic species to test the toxicity of major discharges. If the species survives in the discharge for a certain length of time at certain concentrations, toxicity levels are deemed acceptable.

Another way to test for this group of harmful chemicals is by sampling and examining fish tissue for toxic residue. As a result of these tests, various areas of the state have been closed to fishing, including 60 miles of the Upper Trinity River due to elevated levels of chlordane (a pesticide) and polychlorinated biphenyls (PCBs). PCBs are mixtures of man-made chemicals with similar chemical structures. They are known to cause cancer in animals and are a probable cause of cancer in humans. PCBs were first manufactured in 1929 and were mainly used in power transformers. Once introduced into the food chain, they are stored in the body fat for many years. The EPA banned PCB production in 1977, but residues still exist throughout the United States and even in the polar ice cap.

Petroleum products are organic compounds, many of which are carcinogenic. They often enter groundwater or streams due to mishandling of the drilling process, leaky storage tanks, illegal discharges, and spills during transport. Besides gasoline, diesel, and oil, there are many additives to these that are possibly toxic or carcinogenic. One of these, methyl tertiary butyl ether (MTBE), is a chemical additive to gasoline used since the early 1990s that replaced the function of lead after its ban. Now MTBE is suspected to be a carcinogen in animals, and the EPA is considering a national standard for it. Meanwhile, some states have set their own standards for MTBE in drinking water or banned it altogether. Municipal drinking water supplies in Texas are monitored for MTBE.

The state is investigating various locations that may be contaminated by combinations of chemicals. In the Lower Rio Grande Valley, the Arroyo Colorado, which is an old channel of the Rio Grande, contains excessive amounts of various pesticides, including chlordane and PCBs. Leon Creek near San Antonio and Lake Worth in the Fort Worth area have fish consumption advisories due to PCB contamination. The Houston Ship Channel and Upper Galveston Bay also have a fish advisory due to PCBs.

In 2002 eleven West Texas counties showed elevated levels of perchlorate, a chemical used to make weapons and rockets. The EPA has not established a maximum level for perchlorate, but it is being studied, and the results have shown that excessive levels will affect the thyroid gland, which controls the growth rate in humans.

Chlorine, which is used to eliminate pathogens in drinking water and swimming pools, has been found to form a dangerous by-product, trihalomethanes (THM). THM is thought to cause cancer in humans, and a maximum permitted level has been set by the EPA. Many cities have switched to alternative chemicals, including chlorine dioxide, chloramine, and ozone, to prevent the formation of THM.

As with metals, most wastewater treatment plants do not treat toxins, carcinogens, or organics. As a result they are sometimes discharged untreated into rivers and streams that are the sources of our water supply. Major cities have programs to prevent these chemicals from entering the system in the first place, but, as mentioned, the programs may not be entirely effective.

More and more people are switching to bottled water, but bottled water is even less regulated than municipal water, and the huge number of plastic bottles bring their own environmental issues. Water filter systems are widely used in homes, although the EPA does not endorse any of them. One type of filter, reverse osmosis, forces water through a very fine membrane and removes most contaminants. Much of bottled water is put through a reverse osmosis process. It is not a perfect process and consumes more water than it filters, but many argue that reverse osmosis, especially when coupled with other types of filters, is the best existing filtering process.

Pharmaceuticals and Personal Care Products (PPCP)—These chemicals have just recently begun to be analyzed in our rivers and water supplies. There are thousands of pharmaceuticals and personal care products, and many are discarded, washed off, or passed by humans into wastewater systems that eventually empty into our rivers. There is no agreement on what, if any, effect the various elements of PPCPs have on humans, but there is evidence of altered reproductive functions in some fish due to the level of hormones in the water. The main sources of these hormones are birth control pills and hormone replacement drugs. Another area of concern is the development of small microscopic organisms that are resistant to antibiotics, which have far-reaching negative implications for human health. The EPA has stated that the risks of exposure to PPCPs are basically unknown. As with many pollutants, conventional wastewater treatment systems are not designed to remove these types of chemicals.

Red and Brown Tides and Golden Algae—Algae are often thought of as the harmless green slime that interferes with fishing and swimming. However, there are thousands of species of algae, about a dozen of which are known to cause massive fish kills, contamination of shellfish, and illness in humans. Both harmful and harmless algae can bloom in a variety of colors, but harmful algae can cause serious damage.

Elevated levels of golden algae toxins caused several fish kills in Lake Granbury on the Brazos River in 2005. Photo courtesy of Texas Parks and Wildlife Department.

The Texas coast experiences outbreaks of red and brown tides, and farther inland, in freshwater, there have been golden algae outbreaks. Red tide produces a neurotoxin that affects fish and shellfish. An individual who consumes shellfish affected by a red tide can experience nausea and vomiting. However, consumption of finfish, crabs, and shrimp contaminated by red tide has been shown to be harmless. Red tide can also create airborne toxins that can cause nose, throat, and eye irritation. Brown tide, on the other hand, is not harmful to humans or directly to fish. It does block out sunlight and results in loss of underwater plants, which then reduces the fish population. Golden algae, although harmful to fish, does not pose a risk to humans.

Many coastal states have experienced red and brown tides in the past. Texas had blooms in the 1950s and then again in 1986. Corpus Christi and Aransas Bays had blooms in 1996, followed by a 1997 outbreak in Laguna Madre. In late 2001 another outbreak occurred in Corpus Christi Bay and Laguna Madre. The Texas Parks and Wildlife Department created a task force to study these harmful algae, but there is not yet agreement on the cause of the more frequent outbreaks. Some theories suggest changes in climate and water conditions; others point to pollution and increased nutrients from shore as the root cause.

Golden algae outbreaks occurred in 2005 in Lake Whitney on the Brazos River. There have also been incidents on the Pecos, Colorado, Canadian, Red, and Wichita Rivers. Since 1985 golden algae have killed an estimated 18 million fish, most of which were foreign or rough fish species such as shad and gar.

SOURCES OF POLLUTION
Point Source Pollution

Point source pollution originates from a specific point or location, such as the end of a pipe from a wastewater plant or an industrial discharge. Pollutants from point sources include fecal coliform and nutrients from wastewater and metals and organic compounds from industrial sources. Texas has one of the highest number of point discharges of any state due to its size, its population, and the number of water utilities and petrochemical manufacturing facilities. In 2002 there were 841 industrial, 2,401 municipal, and 578 concentrated animal feeding operation permits in the state. Since 1972 the amount of pollution from point discharges has decreased more than 70 percent, while the amount of waste requiring treatment has increased 85 percent. This improvement can be attributed to better technology and higher required standards of wastewater treatment.

Oil and chemical spills from ships, barges, and offshore oil rigs are point sources of coastal pollution. More than 100,000 ships and barges move 100 million tons of petroleum-related cargo to Texas ports each year. After a couple of spills in the early 1990s, the Texas Oil Spill Prevention and Response Act was passed, and a clean-up fund was established through a per-barrel fee. The largest spill clean-up paid for by the fund was a 5,000-gallon spill in Galveston Bay in 1996.

Abandoned oil wells are another type of point source pollution for which a state fund exists. More than 15,000 abandoned wells have been capped under this program, but thousands more still remain open. It is estimated that 1.5 million holes were drilled for oil and gas in the twentieth century.

Abandoned water wells are a particular concern with regard to groundwater contamination. Since 1965 an estimated 150,000 of the water wells drilled have been abandoned or have deteriorated, thus creating conduits for contaminants to reach groundwater. Over 800,000 water wells were drilled in Texas in the twentieth century.

Nonpoint Source Pollution

Nonpoint source pollution is pollution that cannot be attributed to one specific point or location. It includes runoff from agriculture, mining activities, roads, and parking lots. Septic tanks are also considered a nonpoint source because they are dispersed across a developed landscape and release excess nutrients into the soil. These nutrients may enter the groundwater from many directions, so that the pollution is often not traceable to a spe-

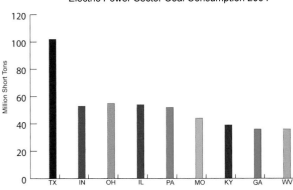

Electric Power Sector Coal Consumption 2004

Top Ten States
Source: Energy Information Administration
Annual Coal Report, DOE/EIA-0584

Texas used the most coal for electricity production of all these states in 2004. Data Source: Texas Water Development Board.

cific tank. Today nonpoint source pollution is the major water quality issue in Texas.

Airborne Pollution in Water

The major airborne pollutant affecting water quality in Texas is mercury, which is emitted from coal-burning power plants. The United States is the first country in the world to regulate mercury emissions from coal-fired power plants. Texas is the nation's number one producer of mercury from power plants, and under new EPA rules the state will have to begin addressing its mercury emissions by 2010. Several lakes in East Texas have mercury advisories recommending limited consumption of certain fish. The entire Texas Gulf Coast is under a mercury advisory to limit consumption of king mackerel. Mercury can cause low intellect, skin tumors, nervous disorders, and birth defects and is especially harmful to pregnant women and children.

HOW IS WATER QUALITY MANAGED IN TEXAS?
Wastewater Permits

All municipal, industrial, and concentrated animal feeding op-

Industry in Texas has a major impact on water quality. Photo courtesy of Texas Parks and Wildlife Department.

eration dischargers must have a state permit from TCEQ, which establishes limits for various pollutants. These limits vary depending on the quality of the water body that receives the discharge. A water body that has been classified by the state as exceptional has stricter standards than one classified as having low aquatic life use.

Water Quality Standards

Texas rivers and streams are divided into segments of varying sizes, each of which has a designated use and specific standards required to maintain that use. For example, a spring-fed segment like the Upper San Marcos River is classified as having exceptional aquatic life habitat, recreational use, including fishing and swimming, and drinking water use. To maintain these uses, there are defined standards for dissolved oxygen, temperature, total dissolved solids, turbidity, bacteria, and other characteristics. Some of these standards, specifically dissolved oxygen, will be higher in segments such as the Upper San Marcos than most segments on other rivers, such as those with an intermediate aquatic life use. Standards are reviewed every three years and are subject to change. Given the vast number of streams in Texas, many are not classified and fall into default standards, unless there is a need for closer examination of the segment. The significance of these standards is that they determine the level of additional contaminants that may be added to the stream.

Section 303d List and TMDLs

Every two years the state has to assess the quality of its water and report to the EPA. The list of stream segments that do not meet their standards is called the 303d list. If a segment is rated in the worst category on the 303d list, the state is required to develop a scientific model called a total maximum daily load, or TMDL. A TMDL shows the amount of a pollutant that the impaired body of water can receive and still maintain its water quality standard. For instance, the TMDL model run on the Arroyo Colorado in the Lower Rio Grande showed the allowable amount of several pollutants was zero, meaning that it could not handle any increase in certain pollutants. By April 2004 TMDLs existed for thirty-three bodies of water in Texas.

Outstanding Natural Resource Waters

Under the federal Clean Water Act, a state is allowed to designate water bodies as Outstanding Natural Resource Waters

(ONRW). This designation provides water bodies with the highest level of protection and prohibits activities that would degrade them, such as wastewater discharges. In 1994 five stream segments, lakes, and bays were proposed as ONRWs, but due to political pressures all were denied. There have been no more proposed ONRWs as of 2005 in Texas, whereas Arkansas has 70, Louisiana has 40, and Oklahoma has 120. An example of a proposed ONRW in Texas is Barton Springs.

Superfund Sites

In 1980 the federal government established what is commonly known as the Superfund to handle the threat of contamination due to releases of toxic chemicals from particular sites that could endanger the public and to help pay for study and cleanup of these sites. One of the most famous Superfund sites is Three Mile Island in Pennsylvania, where a nuclear power plant accident contaminated the surrounding area. One of Texas' Superfund sites is at the ALCOA plant in Lavaca Bay. The funds come from taxes on the chemical and petroleum industries. Often these sites are abandoned, and the responsible parties no longer exist or cannot be identified. If a site is not deemed critical enough for federal assistance, the state has a Superfund program to take care of it.

Stormwater Programs

The EPA requires stormwater management plans for cities with populations of 100,000 or more to deal with pollution that can be carried to streams and rivers via stormwater. Stormwater runoff can contain oil, grease, antifreeze, numerous chemicals, detergents, nutrients, and bacteria. Management plans include identifying points that discharge pollutants, eliminating the discharge, and reducing nonpoint source pollution.

Texas Stream Team

Texas Stream Team, a unit of the River Systems Institute, is a collaborative watershed education and volunteer water quality monitoring program that partners with federal, state, and local stakeholders to improve knowledge and understanding of surface water resources. The program emphasizes a watershed management approach and uses volunteers who implement scientific principles and quality assured methods to collect water quality information from hundreds of sites throughout Texas. Texas Stream Team participants have sampled 880 sites more

A Texas Stream Team monitor collects water quality samples, adding valuable information to a statewide water quality database. Photo courtesy of the River Systems Institute.

than 22,380 times since its inception in 1991. The program is headquartered on the Texas State University at San Marcos campus adjacent to Spring Lake.

THE CLEAN WATER ACT: IS WATER CLEANER?

The 1972 federal Clean Water Act was passed during a time in U.S. history infamous for noticeable pollution in our waterways, including the notorious river fire of 1969, when the Cuyahoga River in Cleveland, Ohio, caught fire due to the abundance of flammable pollutants in the water.

There have been many modifications, expansions, and changes in the Clean Water Act since its passage. Originally the act concentrated on chemical pollution in waterways from point sources. Today the trend is a systemwide approach that deals with both point and nonpoint source pollution and seeks to protect or restore the health of entire watersheds. An example of this is the TMDL program, which utilizes a model for a whole segment of a watershed to calculate the impact of all the sources of pollution within the watershed. Prior to this approach, each permit was analyzed individually, without considering the total load that all of the permitted and nonpermitted uses were putting on the stream segment as a whole.

In many cases the federal government has had to step in to gain the cooperation of states in order to enforce measures of the Clean Water Act. Over the years, states were delegated the authority to issue permits without the requirement of an additional federal permit. Texas was one of the last states to be delegated this responsibility, receiving it only in 1998.

Generally, the water in Texas is cleaner than it was thirty years ago, but keeping water clean is an evolving and continuous endeavor. Rivers do not catch on fire anymore, but many of the problems are cumulative and almost unobservable, except over time with scientific studies or by repeated observations by local citizens. With the creation of more than one thousand new chemicals each year, assessing their potential for harm and then keeping the harmful ones out of water bodies is a daunting task.

As more people move from rural to urban areas, the intense development of expanded cities increases the potential for nonpoint source pollution, which is more difficult to analyze and regulate than are point source discharges.

There will always be intense political pressure to allow new products to enter the market without ample time for analysis of potential harm to water supplies. There will also be pressure to continue to use older successful products that may cause harm to the environment. Citizens need to be aware of these processes and pay attention to the health of their water supplies and watersheds and the effects of their own actions on the watersheds where they live.

According to the U.S. Fish and Wildlife Service, waterfowl hunting in Texas had a total economic impact of $206 million in 2001. Photo courtesy of Texas Parks and Wildlife Department.

10. HOW MUCH IS WATER WORTH?

The concept of what water is worth is very complicated, and there is no agreement on what methods should be used to determine its true value. The price actually paid for water seldom reflects its full value. For many uses of water, such as environmental flows, there is often no price at all—or these uses could be viewed as priceless.

In other situations, government programs subsidize the price of water so that the user does not pay the full price. In the western United States, for instance, since 1902 the federal government has been subsidizing projects to develop water for irrigation. Approximately 85 percent of the total cost of these projects was paid for by all U.S. taxpayers; the users paid fees for the remainder. When consumers, such as farmers in the West, do not see the actual dollar costs of water, there is little incentive for conservation.

In Texas the price of water varies from one area to another. In some cases, the price variations are due to partial subsidies from federal projects or river authorities and lack of competitive markets. One way of calculating the value of water is by figuring the impact on the economy when there is not enough water. An ongoing drought in the South Texas Rio Grande region and the reduced outflow from the rivers of Mexico cost the Lower Rio Grande Valley economy about $1 billion in crop losses and associated impacts over a ten-year period. If Mexico had released the 350,000 acre-feet per year minimum flow that it had agreed to during that time, it would have provided $134 million in business and 4,130 jobs. Calculations like these seldom take account of the value of instream flows in the river and to the estuary. The drought was so severe that at times the Rio Grande stopped flowing to the Gulf of Mexico.

ACTUAL DOLLAR VALUE OF WATER
Water Ranching

"Water ranching" refers to the concept of buying or leasing rural land, mainly for the rights to the groundwater underneath. Most discussions of water ranching in Texas involve the name T. Boone Pickens, who founded Mesa Petroleum. The high-profile entrepreneur controls the rights to groundwater under 150,000 acres in the Texas Panhandle over the Ogallala Aquifer. He plans to market the groundwater to major Texas cities, including Dallas, San Antonio, and El Paso. It is not mere coincidence that Pickens comes out of the oil industry and is switching to water. He believes that eventually water may become as valuable as oil. He is probably right. Today bottled water costs more per gallon than gasoline.

Pickens has rights to about 200,000 acre-feet of water, which could serve about a million people. His estimates of cost are as follows: El Paso, $1,773 per acre-foot; Dallas, $800; and San Antonio, $1,170. The difference in price is due to the distance from his well field. Pipelines could cost from $1 billion to $2 billion. The groundwater conservation district that oversees the area where Pickens holds water rights has certain limitations on pumping. Pickens estimates that his proposed projects will use approximately 10 percent of the district's available groundwater. Under the conservation district rules, the Ogallala in the area of his ranch can only be pumped down by 50 percent over the next fifty years; therefore, he estimates that there will be enough water left in the ground for existing use. However, the Ogallala Aquifer is very slow to recharge, and some experts are concerned about the scope of this project. The concern about too much water being pumped from an aquifer and sold, usually outside the region, has led to "water mining," which implies that more water is being taken out of the aquifer than is recharged and that over time the aquifer will be depleted.

Even with the cost of transfer, the price of water from these projects is still relatively inexpensive, less than 0.004 cent per gallon. Although much of Texas is a dry state prone to extended droughts, the price of water does not reflect the limited supplies. In the state of Washington, water sells for as much as $5,500 per acre-foot compared to Pickens' offer of $1,000 per acre-foot.

The price of irrigated land over the Ogallala is higher than that in nearby ranching areas that is not suitable for farming. Irrigated lands bring $600 to $800 per acre, while ranchlands bring about $250 an acre. Pickens claims that only by selling their groundwater can ranchers compete with the irrigators who are draining the water from beneath their land.

Pickens is not the only water rancher in Texas. The city of El Paso has acquired ranches as far away as 150 miles and plans to use the groundwater beneath them for municipal supply. In Kinney County, near Del Rio and the Rio Grande, water developers are proposing to export groundwater from beneath their land to San Antonio. In Central Texas, entrepreneurs are looking at the Carrizo-Wilcox Aquifer as a water source for the rapidly growing Austin metropolitan area. Local groundwater districts have expressed concern about the depletion of the area aquifers and will have to plan for this use in the future.

Water Pricing and Conservation

MUNICIPALITIES Cities are realizing that by funding conservation measures, they can offset the cost of water even at current prices. San Antonio and many other cities are offering rebates for replacing existing toilets and washing machines with low-water-use models. Water-efficient toilets use up to 66 percent less water and efficient washing machines up to 40 percent less. Cities often give away low-flow shower heads to home owners as well. The San Antonio Water System (SAWS) estimates that the $100 per washer rebate has saved 271 acre-feet of water at $600 per acre-foot. The current price for an acre-foot of water from the Edwards Aquifer in the San Antonio area is over $5,000.

During dry summer months, landscape watering can account for up to 50 percent of water use. El Paso and Austin offer rebates for replacing turf landscaping, such as St. Augustine grass, with water-efficient landscaping. The prices for turf replacement reflect the varying markets for water, both in Texas and throughout the western United States:

- San Antonio—$0.10 per square foot
- El Paso—$1.00 per square foot
- Las Vegas, Nevada—$1.00 per square foot
- Albuquerque, New Mexico—$0.40 per square foot

Austin also offers rebates up to $500 for installing rainwater harvesting systems.

Another method of conservation related to the price of water is by fee structure. Many Texas cities charge more for the more water you use, hoping to create incentives to conserve. Some studies show that this rate structure is not entirely successful, at least among homeowners who only pay attention to the average cost of water. Water rate structures are often difficult to understand; few ratepayers are aware of the cost of the last few gallons used.

AGRICULTURE The dollar value of water plays a major role in agriculture, and in some parts of Texas incentives are in place to reward farmers who reduce their usage. In other areas, farmers are employing water-reducing measures simply because it saves money. For instance, in the High Plains many farmers

The Lady Bird Johnson Wildflower Center's cisterns and storage tanks collect rainwater from 17,000 square feet of roof space and can store more than 60,000 gallons. Photo by Tim Hursley and courtesy of the Water Efficiency Journal.

are upgrading to low-energy precision application center-pivot systems that are up to 98 percent efficient. The cost of these per unit is about $35,000, but savings in operating costs, including pumping, justify the expense.

In the Lower Rio Grande Valley, irrigation is primarily from surface water, which has to be transported by canals and ditches to the user. The ditches are often unlined and uncovered, allowing water to be lost through seepage and evaporation. With losses of up to 30 percent, there is room for substantial water saving through improved infrastructure. The city of Roma paid $2.8 million for irrigation canal improvements and used the saved agricultural water as new municipal water.

Many of these infrastructure improvements are eligible f or below-market loan rates from the Texas Water Development Board. The TWDB also funds up to 75 percent of water-conserving equipment for evaluation or demonstration by water districts.

Center pivot irrigation system with water-saving drop nozzles. Photo by Gregg Eckhardt and courtesy of www.edwardsaquifer.net.

Another market-based strategy to reduce agricultural water consumption is the irrigation suspension program. In dry years farmers would be paid to not irrigate. The Edwards Aquifer Authority estimated that this program would cost $99 per acre-foot. The benefit of irrigation suspension programs is that the cost is incurred during dry years only. Instead of constructing a reservoir that has to be maintained in perpetuity in anticipation of a drought, the water is purchased only when there is a drought. This alleviates not only the expense of infrastructure but also the environmental impacts associated with reservoir construction. Although this solution sounds simple, there can be complicating factors. If the farmer's water was used in a drought by the municipality instead of the farmer, there still might not be enough water available for environmental flows. Water planners need to consider all forms of water use, including streamflow, when opening up water markets.

Leasing State Groundwater Rights

In 2002 and 2003 the state of Texas proposed leasing of ground-water beneath state lands in West Texas to a private entity. This idea was very controversial in part because of the lack of notice and public input on the plan. As a result, in 2003 the lieutenant governor established a senate subcommittee to analyze the concept of leasing state groundwater among other water-related issues.

The committee criticized the lack of public input and the lack of a competitive bidding process. The proposed lease was to a private company, Rio Nuevo, and included several hundred thousand acres at about 20 cents per acre. Other groundwater leases in the area were for 50 cents to $2 per acre. Although Texas had previously leased groundwater beneath state lands, these contracts were between the state and other public entities such as municipalities—not private entities. Furthermore, the groundwater districts in the affected areas were not contacted about the pending state/private contract and the project was not included in the 2007 State Water Plan.

In response to some of the criticisms, the General Land Office and the School Land Board that controls most of the state property in question adopted various new rules relating to the leasing of groundwater under state lands. One of the provisions requires that the rules of the affected groundwater district be followed. Any area not covered by a groundwater district may be required to develop groundwater rules consistent with a groundwater district affected by the project. Also, regional planning groups must be notified of the leases prior to final approval. Proposed leases will also be submitted to the School Land Board for review and comment.

While these additional rules address some of the issues raised concerning the lease of groundwater under state lands, there are still many concerns. There is essentially no method of comment on the effect of these projects on the regional water plans. Although there is a system of review of these contracts by the School Land Board, there is no opportunity to approve or reject the lease. In fall 2007 Rio Nuevo was again negotiating with the General Land Office to lease groundwater under state lands. However, Rio Nuevo will have to approach the affected groundwater districts before any activity can be initiated.

The comparatively low price proposed for leasing of state land for the use of underground water resources shows some of the problems in determining the market value of water. Here a significantly lower price was offered for state-owned water as compared to market prices in the area. Whether or not one supports the idea of an open market for water in Texas, as long as large amounts of water are owned and controlled by the state, it will be difficult to have a competitive open market.

Bottled Water—More Expensive than Gasoline

Given the discussions, negotiations, and litigations over the cost of providing water to the growing state of Texas, it is ironic how

In 2006 Americans spent nearly $11 billion buying 8.25 billion gallons of bottled water—an increase in volume of 9.5 percent from the previous year. Photo from iStock.com.

much people are willing to pay for bottled water. Retail prices for bottled water are from 240 to 10,000 times the cost of average tap water. Total sales of bottled water in the United States have increased from $4 billion a year in 1997 to nearly $11 billion in 2006. Anyone involved in water issues, or community issues in general, has probably seen the often angry reaction to an increase in water rates. Meanwhile, 36 percent of Americans, including lower-income groups, drink bottled water more than once a week. According to the National Resources Defense Council, a five-year supply of bottled water to provide the recommended eight glasses a day would cost a person $1,000. If this same amount came from tap water, it would cost $1.65.

Forty-seven percent of consumers of bottled water claim they drink it because they are concerned about the safety of municipal water, when for just $200 they can buy filters for their home systems. About 40 percent of bottled water is simply municipal water put into a bottle. Some bottled water has undergone additional processing. But regulations for bottled water are often weaker than those for tap water, especially if the bottled water does not cross a state line. Typically, about 90 percent of the cost of a bottle of water is for expenses other than the water—bottling, packaging, shipping, and marketing. In 2007 the city of San Francisco prohibited spending city money on bottled water. Among its concerns was the amount of plastic that was thrown away and the petroleum used to make and ship the bottles.

There seems to be no slowing in the growth of bottled water consumption, which has been about 8 to 10 percent per year. How Americans respond to the growing demand will affect our supplies in the future. In 2005 the state of Michigan ordered a moratorium on new or expanded bottled water operations until a water withdrawal law is passed. Maine is also considering taxing bottled water. In Texas the Ozarka Bottling Company won an important state supreme court case in 1999, which upheld their right to pump groundwater even though it would harm their neighbors' well—this was the first Texas Supreme Court case on the rule of capture since the 1950s. It appears that bottled water sales will continue to grow, and with that growth will come unforeseen impacts, rules, legislation, and litigation.

THE VALUE OF WATER AND THE PRICE OF LAND

Land that has water running through it, next to it, or underneath it is becoming more valuable than the same land in the area without water. In King County, a tract of land with a spring-fed pond sold for 25 percent more than similar tracts without springs. In other areas of South Texas land with water rights sells for $600 per acre compared to $300 per acre for land without water rights.

In a state that often fails to recognize the value of water for the environment, these added values to the price of land reflect an actual dollar value related to nonconsumptive use of water. Rural properties are being bought for hunting, fishing, and other recreational use, not to mention the simple aesthetic enjoyment of having water on one's property. With so little public land in Texas (approximately 2.6%, according to the Trust for Public Land), some city inhabitants are willing to pay extra dollars to have a place in the country with water.

Edwards Aquifer Land Value

In Medina and Uvalde Counties over the Edwards (BFZ) Aquifer, irrigated cropland sells for $3,000 to $4,000 per acre when water rights are included. Compare that to land without water rights, which sells for only $800 to $1,500 per acre. The Edwards Aquifer is probably the most restricted aquifer in Texas, and there are unique rules that affect the value of the groundwater. When the rules were first created for the Edwards Aquifer, owners of qualified irrigation land were allocated 2 acre-feet of water for every acre of land. One acre-foot of water can be leased or sold; the other has to stay with the land. Depending on the price of water and the price of crops, it may or may not be beneficial to lease or sell the extra water. After the extra acre-foot of water is sold, the price of land drops about $1,000 per acre.

View of Lake Austin real estate. Photo by Jason Taylor.

Rio Grande Valley Land Value

In the Lower Rio Grande Valley, most water rights are controlled by an irrigation district and the water rights are tied to the land. If water rights are changed from irrigation to municipal use, the volume of water is cut in half. Water rights have been selling for approximately $800 to $1,000 per acre-foot. Most irrigation districts have policies restricting the transfer of water except when needed by a municipality.

Concho River Land Value

Another area of West Texas showing added value for land with water is along the Concho River in the San Angelo area. Land along the South Concho sells for $2,500 to $4,000 per acre with water rights from the river and $1,000 to $3,500 per acre without water rights. Water rights sold without the land are bringing about $1,000 to $1,500 per acre-foot.

WATER VALUE RELATED TO TRANSFER FROM RURAL TO URBAN AREAS

Rural planning regions are increasingly concerned with the economic impacts on their communities from the transfer of water from rural to urban areas. In fact, the Water Code directs the planning groups to assess the impact of their strategies that move water from rural and agricultural areas. For example, Planning Region L, in Central Texas including San Antonio, compared the value of water per acre-foot to the amount of revenue and number of jobs that would be created. The study estimated that an acre-foot of water for industrial use in 2010 would generate $92,268 worth of business, and commercial water use was estimated to produce $10,819 worth of business per acre-foot. This compared to only $350 per acre-foot of business generated by irrigated agricultural use. The projected number of jobs created also showed great disparity. In 2010 industrial use will create 0.5 job per acre-foot and commercial use is projected to provide 0.21 job per acre-foot. Irrigated agriculture is predicted only to provide 0.0085 job per acre-foot.

The plan also points out that the total amount of water needed by industrial and commercial uses in 2060 is still much lower than the total projected for agricultural use. In 2060 projections for industrial use are 179,715 acre-feet per year and for commercial use 159,309 acre-feet per year. Irrigated agriculture is projected to still need 301,679 acre-feet in 2060 in Region L.

The significant variance in the dollar value of water for urban use versus rural use highlights some of the complexities of attempting to apply market-based principles to the price of water. As these values become further apart there will be stress on the rural communities and their way of life. Difficult questions will need to be answered:

- Since the value of computer chips is much higher than corn, should significant amounts of water be moved to the sector of the economy that manufactures computer chips?
- What will the long-term effects of the transfer of water be to streamflows and water quality?
- If the price of food is subsidized by federal programs, how should that figure into the equation of relative values?
- Are we willing to allow whoever pays the most for the water to be able to obtain the water?

COST OF STATE WATER PLAN SOLUTIONS

The cost of the fourteen proposed reservoirs in the state water plan is $5 billion, but there are numerous other costs involved in solving the proposed water shortage in Texas. The state water plan actually estimates the total cost of making up the water shortage by 2060 at $173 billion. The plan breaks down the costs into five categories:

Water supply treatment and distribution	$79 billion
Implementation of water strategies in Water Plan	$30.7 billion
Treatment and collection	$59.1 billion
Flood control	$4.2 billion
Assistance to municipalities to implement Water Plan	$2.1 billion

Cost of Not Planning for Future Water

If projected water needs during a record drought in 2060 are not met, the estimated cost to businesses and workers would be $98 billion. This loss would also be felt by state and local government in the form of reduced taxes, estimated to be $5.4 billion in 2060.

Desalination

Desalination, or the removal of salt from brackish groundwater

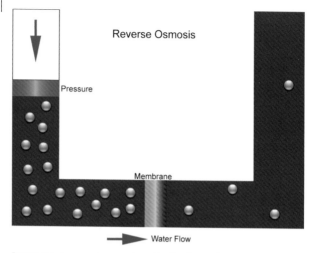

Reverse osmosis process. Image courtesy of the River Systems Institute.

or seawater, is often discussed as the ultimate long-term solution to water supply in Texas and throughout the western United States. Brackish water generally refers to water with more than 1000 mg/liter of total dissolved solids (TDS). For comparison, seawater contains about 35,000 mg/liter of TDS. In West Texas one entrepreneur is using brackish groundwater to raise shrimp, but most users need water that has a TDS of less than 500 mg/liter. Generally this is accomplished by desalination through reverse osmosis. Although the price of reverse osmosis has dropped with new technology and increased production of equipment, the price of desalinated water is often cost prohibitive under any near-term plans. General estimates of the cost of desalination are from $768 to $1,390 per acre-foot for seawater and $429 to $953 per acre-foot for brackish groundwater. This does not include the cost of distribution, which can be a substantial portion of the overhead.

Another cost of reverse osmosis is the disposal of the even more brackish water that is the result of the process. The price of this can vary according to the disposal method, which can include deep well injection, evaporation ponds, and discharge into the sea.

All of these methods have potential environmental impacts on groundwater, surface water, or, in the case of discharge into the sea, a localized impact around the point of discharge. The cost of dealing with the discharge can also depend on the geology of the area and the chemical composition of the discharge (e.g., if there is arsenic or mercury concentrate). Brackish reverse osmosis plants are about 75 percent efficient, meaning they lose 25 percent of their original volume of water to brackish waste. Saltwater plants lose about 70 percent of their original water volume. This means that more raw water is needed than the final usable amount, resulting in oversizing pipes, storage facilities, and even increased cost of raw water.

In the 2007 State Water Plan, eight regional planning areas called for desalination projects with the capacity of an additional 312,000 acre-feet of water. Currently in Texas desalination capacity is 25,570 acre-feet. Worldwide there are more than 12,500 desalination plants in 120 countries, 60 percent of which are in the Middle East. Forecasts show the world market for desalinated water will grow by more that $70 billion in the next twenty years.

WATER MARKETS

The public sector, whether it is river authorities, municipalities, groundwater districts, or water improvement districts, dominate water marketing in Texas. According to some economists, the only active competitive water markets are in the Rio Grande Valley and the Edwards Aquifer, but these have special conditions. Supporters of a more open water market in Texas claim that

- active markets would provide water for growing cities without the cost of infrastructure such as reservoirs;
- water could be purchased from agriculture interests for municipal uses;
- water for drought contingencies could be purchased from agriculture only in dry years;

- active markets can provide water for recreational use and environmental use as recreation and tourism revenues exceed the value of agricultural revenues; and
- markets promote efficient use of water and conservation as costs rise.

Critics of a total market-based water system are concerned that the complete value of water will not be represented, specifically, environmental values. Although recreational values can be calculated to some extent, other environmental values are more difficult to calculate and are not recognized in conventional markets. These values are often hidden and are not reflected in the price of water. Some of these hidden values are as follows:

- Impact on water quality of the remaining water in rivers as more water is diverted and there is less water volume to dilute contaminants.
- Environmental flows for preserving habitat.
- Ecosystem protection, including the estuaries downstream of projects, such as reservoirs.
- Aesthetic value of rivers, streams, and bays, including heritage values and community identity.

When these hidden values are not reflected in the dollar value of water, the markets do not truly reflect the actual value. The result can be trading of water from rural use of less economic value to urban use, without regard for the environment or the rural way of life.

THE HIDDEN VALUE OF WATER FOR THE ENVIRONMENT

Currently in Texas no one actually pays a bill for water to flow in rivers and into estuaries. The persons who use the rivers and bays for recreation, the commercial fishermen who work the waters, the people who consume the fish and shellfish, the cities upstream who depend on the dilution of wastewater, the industries that use saltwater for manufacturing processes, the towns that are protected from hurricanes by the wetlands, the birders who enjoy the waterfowl habitat, the ocean liners, tankers, and barges that use the water for navigation, and those who simply enjoy the view of the bays and estuaries from the shore—all these individuals do not pay a fee for the use of this water. Because no one pays, there is no certificate of ownership that guarantees that water for rivers and estuaries will continue to flow. There are basically no permits for these flows in Texas that can be traded, and therefore water markets do not include values for these uses. A municipality, for example, owns a water right usually in perpetuity that can be bought or sold. This right, depending on the seniority date, provides a type of guarantee that the water will be available under most conditions.

Through this process of water permits, market values are created, and for the permitted water, some of the benefits of water marketing are present. One of the important aspects of water marketing is that as the value increases, users tend to conserve and reduce usage. Without a market there is little incentive to conserve other than concern for the environment or the overall long-term effects on the community. If water for rivers and bays and estuaries had a market value, then water projects would include a cost/benefit analysis comparing the value of the diversion of water to the reduction in the quantity and/or quality of flows for rivers and estuaries. The problem is that not only are there no permits or certificates for instream and freshwater inflows; there is no recognized method of determining the potential values of the water.

The Economic Value of Estuaries

Only since the 1970s have there been concerted efforts to determine the economic value of estuaries, which require freshwater to function. Robert Costanza, a well-known environmental economist, established a value for the services provided by one acre of an estuary at $11,000 per acre per year. These services include all of the hidden values mentioned above, such as dilution of wastewater and hurricane and storm buffering. Applying this number to San Antonio Bay at the mouth of the Guadalupe

The Gulf of Mexico is the most productive shrimp-producing region in the United States. According to the Federal Reserve Bank of Dallas, Texas had 35.2 million pounds of shrimp landings with a value of $96.3 million in 2003. Photo courtesy of Texas Parks and Wildlife Department.

River with 130,000 acres, the value would be almost $1.4 billion per year. When this value is compared to projects that would significantly reduce the freshwater flows to the bay, many projects would not be able to compete with the value of the bay.

Many people do not recognize the environmental values as part of market value, so other methods of calculating the value of a bay have been performed. The Texas Parks and Wildlife Department analyzed San Antonio Bay looking at the commercial fish and shellfish harvest, recreational fishing, birdwatching, and other recreational activities. TPWD determined that the bay had a value of $55 million per year, which, although much less than Costanza's estimates, would still be of significant enough value to render some competing projects unaffordable. For example, the city of San Antonio and the Guadalupe Blanco River Authority considered a project that would reduce the optimum flows to the estuary and result in a 40 percent reduction in productivity of San Antonio Bay, according to the state's model. Using the TPWD value, this 40 percent reduction would cost $22 million per year. Adding this $22 million loss in value of the bay to the cost of the water for the Lower Guadalupe Diversion

Project would increase the acre-foot per year cost from $800 to $4,600. Realizing that these are general estimates, the process itself shows how applying values for environmental flows has the potential to reduce the apparent economic benefit of some water development plans.

Recreational Values of Water

Probably the most progress in calculating values for instream and freshwater inflow is in the area of recreational use. This success may just be related to the fact that the values are more recognizable by the majority of the public who themselves participate in water-related recreation. It is not easy to understand some of the more complex values of estuaries. For example, although phytoplankton is important and forms the basis of the food chain, the average person is not familiar with its role in the economy. The U.S. Fish and Wildlife Service conducted a study in 1996 that found that 76 percent of Americans, or approximately 171 million people, are involved in recreational use of water. Sixty-five million of these are fishermen who spend about $626 million a year on equipment and travel. The total value to the economy of water for recreation is estimated at $100 billion

There are currently seven official coastal and one inland paddling trails in Texas. Photo courtesy of Texas Parks and Wildlife Department.

a year, including $25 billion for boating equipment. According to TPWD, recreational anglers in Texas spend about $2 billion a year on fishing trips and gear. The state ranks fifth in the nation for boater registrations and second in saltwater fishing among coastal states.

The U.S. Forest Service (USFS) turned some of these large recreation numbers into dollars per visitor day in the forests and arrived at the following:

Activity	Value in Dollars per Visitor Day
Swimming	$18.92
Hunting	$72.59
Fishing	$134.47
Wildlife viewing	$95.00
Scenery	$14.35

The comparative value of these uses is interesting. Note that fishing is the highest value and far outranks swimming. Obviously, a lot of the value is related to the amount of equipment involved but also the income of the people who participate. Interestingly, wildlife viewing ranks higher than hunting. For example, serious birders spend more than $1,000 per year on equipment alone. It is also interesting to see scenery included as an economic concept and as a recreational use, although it has the least value.

Other Hidden Values of Water

A German study on the economic value of the Rhine reveals some interesting value considerations for a flowing river:

Function of the Rhine River	Value per Year
Clean drinking water	$663 million
Fish production	$1.7 million
Existence value of nature	$640 million
Natural retention capacity	$500 million
TOTAL	$1,800 million

The most notable comparative value is the second ranked Existence Value of Nature, which almost equals the value of Clean Drinking Water. The comparatively low value of fish production is also noteworthy. This study is mentioned not necessarily to prove the actual worth of these values but rather to illustrate that these concepts do exist and should be analyzed when comparing the value of any water development projects.

In New Mexico a controversy over instream flows in the Rio Grande near Albuquerque is forcing the issue of the value of water for various uses, including survival of endangered species. A federal court has ordered a minimum flow requirement for a portion of the Rio Grande to support the habitat of the silvery minnow. Maintaining these minimum flows will require the city of Albuquerque and farmers along the Middle Rio Grande Conservancy District to reduce their diversions from the river. Critics point out the economic burden of not being able to use this water for municipal and agricultural use.

The result has been several analyses of the value of the water and the cost of not being able to utilize it. One group of opponents claim that 48,200 acre-feet of upstream water rights valued today at $4,500 per acre-foot would have to be replaced. This would mean a cost to the city of Albuquerque of over $300 million, including transaction costs. Another study examined the cost of lost agricultural production and jobs in the Middle Rio Grande area. Since the minimum flow requirements only affect drier years, the economic effects were apportioned across the dry years where additional flow would be necessary. The study estimated a loss of about $6 million per year in agricultural production along with 362 jobs. This represents less than 0.1 percent of the annual output of the Middle Rio Grande area.

Finding the true cost of water is a complicated task. As testimony to this complexity, another study of the impact of water for the silvery minnow calculated the downstream benefits of having additional flow that was formerly diverted. According to the study, southern New Mexico would gain $217,000 per year from the additional water left in the Rio Grande for the silvery minnow. Farther downstream, near El Paso, the extra water

The Rio Grande near Taos, New Mexico. Photo by Jerry Kimmel.

would provide farmers with an additional $203,000 per year and the city of El Paso with $1.27 million annually. Additional environmental flows would provide positive economic benefit even for water users in the Upper Rio Grande Basin in southern Colorado and miles downstream of El Paso in Fort Quitman.

It is important to note that additional values of water downstream of protected areas are due in part to the assumption that much of the protected water will be consumed downstream. This illustrates the sad fact that the preservation of environmental flows in one area does not necessarily mean that those flows will stay in the stream past the protected area or reach the estuary. Unfortunately, some methods of evaluating instream flow reflect the downstream gains where flows are not protected for the environment. It could be argued that adding the value of consuming water downstream to the value of preserving the water upstream gives a false reading. These offsetting values do not necessarily reflect the true value of environmental flows, although in some cases the result would show an increased water value.

THE VALUE OF THE SYSTEM

We have discussed values for rivers and estuaries, in terms of both dollar values and hidden values. Estuaries are some of the most productive places on Earth, but the water they need comes from the rivers; therefore, the value of water flowing in a river is not just for the river itself but also for the estuary it feeds. Most models show that the total flows needed for estuaries are greater than those needed for the rivers that feed them. Rivers need certain flows more instantaneously, while estuaries require flows by seasonal volume. Estuaries cover just 6 percent of the Earth's surface, but the ecosystem services they perform comprise 34 percent of the economic value of the world. Ecosystem services include many of the previously discussed functions of an estuary, including dilution of pollutants, providing habitat, and flood and storm reductions. Ecosystem services are not just confined to estuaries but to rivers, forests, and even the air. Estimates are that the value of all of the Earth's ecosystem services exceeds three times the global gross national product, or the measure of the total market value of all goods and services. Perhaps as the awareness of ecosystem value grows, policies will be put in place that weigh the full impact of major projects on the water resources before they are allocated.

A fisherman poles his kayak across a Texas bay. Photo courtesy of Texas Parks and Wildlife Department.

11. WATER IS OUR LEGACY

Just a few short weeks after the New Year in 2001, the once-mighty Rio Grande, draining a 175,000-square-mile basin and traveling nearly 2,000 miles, stopped flowing just yards before its entrance to the Gulf of Mexico. At one time, the Rio Grande, which is the world's twenty-fourth longest river, was wide enough at the mouth that it could accommodate oceangoing ships. In Brownsville the river was more than 100 feet wide just a few decades ago. Today it may be less than 20 feet across. This sad situation is the result of a number of factors all working against the sustainability of this vital source of freshwater. The river has historically served millions of people in the United States and Mexico, both uniting them and dividing them physically and politically. Nearly a decade of drought, choking even the hardiest exotic aquatic weeds, and increasing use of water for farms and municipalities finally took its toll. The magnitude of the drought's impact was felt well beyond the borders of both nations.

In the years since those bleak images first appeared in the media around the world, the Rio Grande has become a poster child for the threatened sustainability of surface water in Texas. The stark reality that such a seemingly boundless resource could actually be exhausted was, like the drought of the 1950s, another wake-up call: action is needed now to sustain Texas' freshwater resources, particularly as global warming accentuates drought, evaporation, and demand for water itself.

First and foremost, we need to strengthen our ability to manage entire watersheds as unified systems, regardless of state or national boundaries.

Deforestation, or, conversely, the invasion of exotic tree species such as tamarisk and mesquite, in one end of a river basin can have major effects on the availability of water in the other. The spread of impervious cover, which accelerates runoff and impedes aquifer recharge upstream, can dramatically change conditions for interests downstream. Today virtually all of Texas' rivers are managed systems as the result of the extensive series of reservoirs and other infrastructure built after the 1950s. Management must extend to all aspects of watershed function in ways that view rivers as systems that depend on the sum of their parts—aquifers, drainage areas, streambeds, and estuaries. Everyone has a role to play.

In Texas virtually all land in the watersheds is private property. Therefore, the stewardship of those lands by private citizens, not government, will be the determining factor in how well they contribute to the sustainability of the rivers and streams they help to supply. Most surface and groundwater supplies in Texas have their origin in rainfall that is collected in a complex system of geology, soil, and vegetation. When this system is managed properly, flooding is minimized, aquifers are replenished, and water flows steadily into the rivers and lakes where it can be used by humans or the environment. Consequently, like the landscape itself, much of the future of our water supplies lies in the actions of private citizens on properties they own and control.

Thus private land stewardship must be part of the water

Devils River State Natural Area. Photo courtesy of Texas Parks and Wildlife Department.

equation in Texas. Public policy should be designed and implemented to support, encourage, and reinforce good behavior and
permanent protection of the landscape. One tool used widely in
other states, but still in its infancy in Texas, is the purchase of
development rights—where a landowner sells a conservation
easement over his or her property and protects it in perpetuity
from adverse use but continued private ownership and use of
the land itself are allowed. Organizations such as the American
Farmland Trust and the Texas Land Trust Council have been
at the forefront of efforts to establish a purchase of development rights program whereby rights are purchased to protect
the property from being subdivided. The program would keep
property values down and provide family farmers and ranchers
with a means of staying on the land. The public benefits of this
kind of arrangement are protection of open space, watershed
function, water recharge, and wildlife habitat. The city of San
Antonio has successfully implemented a purchase of development rights program to protect recharge areas of the Edwards
Aquifer.

Short of purchasing development rights or donating conservation easements, which accomplish permanent protection, the federal government provides assistance to private landowners who desire to preserve important natural habitat or other natural values on their property. Funding for land conservation is available by the U.S. Department of Agriculture through the Federal Farm Bill. However, efforts must be initiated by the individual landowner.

Another important function of the private sector in meeting our water needs for the future is the establishment of a water market. For too long, we have treated water as if it were free and not a commodity with value like petroleum or corn. This has had the unfortunate result of both encouraging irresponsible use of water and inhibiting its transfer from one water right holder to another. For many, the prospect of water becoming a commodity is uncomfortable because it has always been viewed as a public resource. Yet we have transferred virtually all of the publicly owned surface water to interests that use it for economic purposes, including agriculture, and to interests that will be reluctant to transfer it for other uses without compensation.

Moreover, it may be difficult to bring substantial amounts of groundwater to bear on our future water needs without creating a marketplace in which entrepreneurs can both invest in that objective and expect a profit for doing so. Finally, as we have seen with gasoline in recent years, rising costs have been the greatest incentive spurring the purchase of more fuel-efficient automobiles, and we could expect the same investment in water efficiency to occur if it made more economic sense. We need to accept the fact that a major component of conserving our water resources in Texas will be market-based acquisitions of surface water and groundwater rights.

We must find a way to move water westward in Texas. This is inevitable, but, again, the prospect of interbasin transfer is one that is not universally accepted. The fact is that we are already transferring water from one basin to another in a number of relatively small instances around the state, and such transfers may also give us the opportunity to accomplish environmental goals.

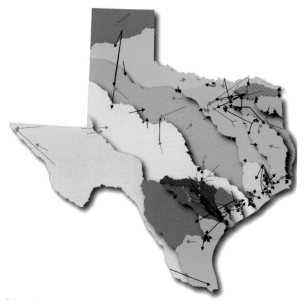

Existing (black) and proposed (red) interbasin transfers from the 2002 Water for Texas.
Data Source: Texas Water Development Board.

One reason that we are having real problems providing sufficient environmental flows in our rivers and streams is that in many of them, all of the water has already been appropriated by the state for other uses. It is too late to regulate. Approval of interbasin transfers on any large scale by the state and federal governments will require project sponsors to meet numerous environmental requirements and mitigate environmental impacts. Soon a condition of approval of such projects will be the requirement that water be guaranteed for environmental flows. The bottom line is that the need is sufficiently great in the western part of the state and there is so much water in the east that it is going to happen. It is up to us and to policy makers to produce gains for the environment along the way.

If we do not address the serious environmental issues associated with our growing demand for water, the development of new water resources itself will be jeopardized. There was a time in Texas when every single major water project, including reservoir construction and increased pumping from the Edwards Aquifer, was sure to be fought out in federal court. During the 1990s, this battlefield atmosphere was replaced by a consensus approach to water planning that brought all voices and stakeholders to the table and into the process.

Today we are moving beyond theory and implementing real water projects, which can be measured in gallons and acre-feet. The prospect of a return to the courthouse is now very real. This approach is costly to everyone and often does not produce positive results. Unless we can successfully implement some of the significant agreements that have been forged between environmentalists and the water development community, the courts may start deciding for us again.

The recreational value of our rivers and streams must also be protected and enhanced. Historically, both the state government and recreational river users have resisted funding organized efforts to create additional recreational opportunities in our rivers. An exception to this has been the installation of boat ramps. Sadly, there have been no corresponding recreational infrastructure investments in river basins where the physical features prohibit motorboat use even though they may provide wonderful opportunities for fly fishing, canoeing, and kayaking. One method of providing funds to support rivers could be to establish a fee for paddling, just as we have fees for motorboating and fishing. This revenue would enable the state to begin investing in provisions for nonmotorized recreational use of our rivers and greatly build the constituency for protecting them, a constituency that today has all too small a voice in the development of water policy in Texas.

Another constituency whose voice must be strengthened is that of the groundwater management community. Funding must be found to increase the scientific capacity, technical resources, and management expertise of the newly emerging groundwater conservation districts. It is going to be very difficult for them to do their jobs with the limited technical support they possess today concerning the resources they are empowered to protect.

Krause Springs in Spicewood, Texas. Photo courtesy of Texas Parks and Wildlife Department.

To compound the problem, their formation largely along county boundaries, rather than around the natural configuration of the aquifers they manage, potentially puts them at odds with neighboring districts that share the same underground water resource.

Shared resources are an even greater policy issue with respect to the marked disconnect between groundwater and

Learning to fish at Garner State Park. Photo courtesy of Texas Parks and Wildlife Department.

surface water management. There is no longer any scientific question that groundwater resources are linked to surface water. Overpumping can make springs go dry and thus has a significant impact on the rivers and streams into which the springs feed. To continue to pretend that aquifer use above the headwaters of our rivers and their tributaries is a different issue from withdrawal of water from the streams themselves is to risk major damage to our river systems. Policy makers must push for conjunctive use of these resources, treating them as they actually are: the same.

We need improvements in water quality protection as well. Since the establishment of the U.S. Environmental Protection Agency in the 1970s, we have made great progress in water quality by regulating point sources of pollutants. Today most water quality degradation occurs as the result of nonpoint source pollution from runoff of urban streets, parking lots, and agricultural operations that use pesticides, herbicides, and fertilizers. This is a much more difficult problem to address because the discharges do not come from single locations, which can be controlled. Further, water quality management historically has been reactive and implemented only when a water body has already become impaired. It is imperative that we educate watershed landowners and encourage voluntary measures to limit the impact of agricultural chemicals and urban runoff before water quality becomes a problem.

In the final analysis, education and engagement hold the best hope of getting us through what will surely be the greatest natural resource management challenge of our time—making sure we have enough water for our future and protecting vital supplies for the environment. Many people still try to make a distinction between these two purposes as if they were at odds. It is not uncommon to hear water experts speak of water for people and water for wildlife, implying that water for the environment requires a sacrifice of water for human use. Anyone who still believes this ignores the reality that saltwater fishing alone is a billion-dollar business on the Texas Gulf Coast and is dependent on continued supplies of freshwater to bays and estuaries.

Sadly, this and so many other facts about water are unknown to most of our citizens, who thus tend toward apathy and disinterest. We can change that. The most important thing that each and every one of us can do to help provide a future water supply and quality of life for our children is to take the time to get them involved. Today in Texas most citizens are growing up in urban settings that provide little contact with the outdoors. Demographically, the least likely Texans to fish or canoe or find recreation in the rivers, streams, and lakes are people of color, who constitute a majority of the population. Thus it is imperative that we do everything possible to introduce children of all races and backgrounds to the wonderful opportunities that our water resources provide us.

A child who loves to fish will become an adult who will work to protect our fisheries. A child who loves to canoe will become an adult who will fight to protect our rivers. We will fail them miserably if we do not make the effort to engage them in both the joys and the responsibilities of using and caring for our water resources, for they are the voters and taxpayers of the future. More important, if children grow up without the opportunities we have had in our lifetimes to experience the spectacular aquatic environment of our state, they will miss one of the greatest joys and privileges of being Texans.

APPENDIX

Contact Information for Texas Water Entitites

STATE AGENCIES AND ENTITIES

Texas Commission on Environmental Quality

Texas Commission on
Environmental Quality Water
Supply Division
12100 Park 35 Circle
Austin, TX 78753
512-239-1000
www.tceq.state.tx.us/

Rio Grande Watermaster Program

Rio Grande Watermaster
1804 West Jefferson Avenue
Harlingen, TX 78550-5247
956-425-6010
www.tceq.state.tx.us/
compliance/field_ops/
wmaster/rgwr/riogrande.html

South Texas Watermaster Program

South Texas Watermaster
14250 Judson Road
San Antonio, TX 78233-4480
800-733-2733
www.tceq.state.tx.us/
compliance/field_ops/
wmaster/stwr/southtexas.html

Texas General Land Office

Texas General Land Office
1700 North Congress Avenue
P.O. Box 12873
Austin, TX 78711-2873
800-998-4GLO
www.glo.state.tx.us/

Texas Parks and Wildlife Department

Texas Parks and Wildlife Department
4200 Smith School Road
Austin, TX 78744
800-792-1112
512-389-4800
www.tpwd.state.tx.us/

Texas Water Development Board

Texas Water Development Board
1700 N. Congress Avenue
P.O. Box 13231
Austin, TX 78711-3231
512-463-7847
www.twdb.state.tx.us/home/index.asp

Texas Railroad Commission

Railroad Commission of Texas
1701 North Congress Avenue
P.O. Box 12967
Austin, TX 78711-2967
General information
512-463-7288
www.rrc.state.tx.us/

Regional Water Planning Groups

REGION A—PANHANDLE WATER PLANNING AREA

Panhandle Regional Planning Commission
P.O. Box 9257
Amarillo, TX 79105
806-372-3381
www.panhandlewater.org/

REGION B—RED RIVER REGIONAL WATER PLANNING GROUP

Red River Authority of Texas
3000 Hammon Road
P.O. Box 240
Wichita Falls, TX 76307-0240
940-723-8697
www.rra.dst.tx.us/water_planning.htm

REGION C—WATER PLANNING GROUP

Region C Water Planning Group for North Texas
5525 N. MacArthur Boulevard, Suite 530
Irving, TX 75038
972-442-5405
www.regioncwater.org/index.cfm

REGION D— NORTH EAST TEXAS WATER PLANNING GROUP

Texas Water Development Board
1700 N. Congress Avenue
P.O. Box 13231
Austin, TX 78711-3231
903-639-7538
www.twdb.state.tx.us/publications/newsletters/e-news/rwpgmeetings.htm

REGION E—FAR WEST TEXAS WATER PLANNING GROUP

Far West Texas Water Planning Group
1100 North Stanton, Suite 610
El Paso, TX 79902
915-533-0998
www.riocog.org/EnvSvcs/FWTWPG/fwtwpg.htm

REGION F—WATER PLANNING GROUP

Region F Water Planning Group
P.O. Box 869
Big Spring, TX 79720
432-267-6341
www.crmwd.org/

REGION G—BRAZOS WATER PLANNING GROUP

Region G Brazos River Authority
P.O. Box 7555
Waco, TX 76714
254-761-3177
http://www.brazosgwater.org/2.html

REGION H—WATER PLANNING GROUP
Texas Water Development Board
1700 N. Congress Avenue
P.O. Box 13231
Austin, TX 78711-3231
512-463-7847
www.twdb.state.tx.us/
rwpg/mtg-updates/h.htm

REGION I—EAST TEXAS REGIONAL WATER PLANNING GROUP
East Texas Regional Water Planning Group
210 Premier Drive
Jasper, TX 75951
409-384-5704
www.detcog.org/etrwpg/

REGION J—PLATEAU WATER PLANNING GROUP
Texas Water Development Board
1700 N. Congress Avenue
P.O. Box 13231
Austin, TX 78711-3231
512-463-7847
www.twdb.state.tx.us/
rwpg/mtg-updates/h.htm

REGION K—LOWER COLORADO REGIONAL WATER PLANNING GROUP
Texas Water Development Board
1700 N. Congress Avenue
P.O. Box 13231
Austin, TX 78711-3231
512-463-7847
www.twdb.state.tx.us/home/
index.asp

REGION L—SOUTH CENTRAL TEXAS REGIONAL WATER PLANNING GROUP
South Central Texas Regional Water Planning Group
111 W. 8th Street
Austin, TX 78701
512-583-0929
www.watershedexperience
.com

REGION M—RIO GRANDE REGIONAL WATER PLANNING GROUP
Lower Rio Grande Valley Development Council
311 N. 15th Street
McAllen, TX 78501-4705
956-682-3481
www.riograndewaterplan.org/

REGION N—COASTAL BEND REGIONAL WATER PLANNING GROUP
Coastal Bend Regional Water Planning Group
c/o Nueces River Authority—Coastal Bend Division
6300 Ocean Drive, Unit 5865
Corpus Christi, TX 78412-5865
361-825-3193
www.nueces-ra.org/CP/RWPG/

REGION O— LLANO ESTACADO REGIONAL WATER PLANNING GROUP
Llano Estacado Regional Water Planning Group
2930 Avenue Q
Lubbock, TX 79411-2499
806-762-0181
www.llanoplan.org/

REGION P—LAVACA WATER PLANNING GROUP
LNRA 4631 FM 313
P.O. Box 429
Edna, TX 77957
361-782-5229
www.lnra.org

River Authorities

ANGELINA AND NECHES RIVER AUTHORITY
ANRA Main Office
210 East Lufkin Avenue
P.O. Box 387
Lufkin, TX 75901
936-632-7795
www.anra.org/

BRAZOS RIVER AUTHORITY
Brazos River Authority
4600 Cobbs Drive
P.O. Box 7555
Waco, TX 76714
254-761-3100
Fax 254-761-3207
www.brazos.org/

GUADALUPE-BLANCO RIVER AUTHORITY
Guadalupe-Blanco River
Authority
933 East Court Street
Seguin, TX 78155
830-379-5822
www.gbra.org/

LAVACA-NAVIDAD RIVER AUTHORITY
Lavaca-Navidad River
Authority
4631 FM 3131
P.O. Box 429
Edna, TX 77957
361-782-5229
www.lnra.org/

REGION K—LOWER COLORADO RIVER AUTHORITY
LCRA Headquarters
3700 Lake Austin Boulevard
Austin, TX 78703
512-473-3200
1-800-776-5272
www.lcra.org/

LOWER NECHES VALLEY AUTHORITY
Lower Neches Valley Authority
7850 Eastex Freeway
Beaumont, TX 77708
409-892-4011
www.lnva.dst.tx.us/

NUECES RIVER AUTHORITY
Nueces River Authority
General Office
First State Bank Bldg, Suite 206
200 E. Nopal - P.O. Box 349
Uvalde, TX 78802-0349

COASTAL BEND DIVISION
Natural Resource Center
6300 Ocean Drive Unit 5865
Corpus Christi, TX 78412-5865
830-278-6810
361-825-3193
www.nueces-ra.org/

RED RIVER AUTHORITY
Red River Authority of Texas
3000 Hammon Road
Wichita Falls, TX 76310
940-723-2236
www.rra.dst.tx.us/

SABINE RIVER AUTHORITY
Sabine River Authority of
Texas
P.O. Box 579
Orange, TX 77631-0579
409-746-2192
www.sra.dst.tx.us/

SAN ANTONIO RIVER AUTHORITY
San Antonio River Authority
100 East Guenther Street
San Antonio, TX 78204
210-227-1373
www.sara-tx.org/

TRINITY RIVER AUTHORITY
Trinity River Authority of
Texas
P.O. Box 60
Arlington, TX 76004
817-467-4343
www.trinityra.org/

UPPER COLORADO RIVER AUTHORITY
Upper Colorado River
Authority
412 Orient
San Angelo, TX 76903
325-655-0565
www.ucratx.org/

UPPER GUADALUPE RIVER AUTHORITY
Upper Guadalupe River
Authority
125 Lehmann Drive, Suite 100
Kerrville, TX 78028
830-896-5445
www.ugra.org/

FEDERAL AGENCIES AND ENTITIES

U.S. Army Corps of Engineers
U.S. Army Corps of Engineers
Fort Worth District
P.O. Box 17300
Fort Worth, TX 76102
817-886-1326
www.swf.usace.army.mil/

U.S. Department of Agriculture
U.S. Department of Agriculture
1400 Independence Ave. S.W.
Washington, DC 20250
www.usda.gov

Natural Resource Conservation Service
USDA-Natural Resources Conservation Service
101 South Main
Temple, TX 76501
254-742-9800
www.tx.nrcs.usda.gov/

U.S. Forest Service
National Forests & Grasslands
in Texas
415 S. First Street, Suite 110
Lufkin, TX 75901
936-639-8501
www.fs.fed.us/r8/texas/

U.S. Department of the Interior
Department of the Interior
1849 C Street, N.W.
Washington, DC 20240
202-208-3100
www.doi.gov/

National Park Service
Southeast Region
National Park Service
100 Alabama Street S.W.
1924 Building
Atlanta, GA 30303
404-562-3100
www.nps.gov/state/tx/

U.S. Bureau of Reclamation
U.S. Bureau of Reclamation
Great Plains Regional Office
P.O. Box 36900
Billings, MT 59107-6900
406-247-7600
www.usbr.gov/gp/

U.S. Fish and Wildlife Service
U.S. Fish and Wildlife Service
Southwest Region
P.O. Box 1306
Albuquerque, NM 87103-1306
505-248-6911
www.fws.gov/southwest/
statelinks/texaslinks.htm

U.S. Environmental Protection Agency
U.S. Environmental Protection
Agency
Region VI
1445 Ross Avenue, Suite 1200
Dallas, Texas 75202
214-665-6444
www.epa.gov/region6/

U.S. Geological Survey
U.S. Geological Survey
Texas Water Science Center
8027 Exchange Drive
Austin, TX 78754-4733
512-927-3500
http://tx.usgs.gov/

U.S. International Boundary and

Water Commission

U.S. International Boundary
and Water Commission
4171 North Mesa, Suite C-100
El Paso, TX 79902
915-832-4157
www.ibwc.state.gov/

Federal Emergency Management Agency

Federal Emergency Management Agency
500 C Street, S.W.
Washington, DC 20472
202-646-4006
www.fema.gov/

NONPROFIT ORGANIZATIONS

Caddo Lake Institute

Caddo Lake Institute
44 East Avenue, Suite 100
Austin, TX 78704
512-482-9345
www.caddolakeinstitute.us/

Galveston Bay Conservation and Preservation Association

Galveston Bay Conservation
and Preservation Association
P.O. Box 323
Seabrook, TX 77586
281-291-8355
www.gbcpa.net/

Greater Edwards Aquifer Alliance

Greater Edwards Aquifer
Alliance
1809 Blanco Road
San Antonio, TX 78212
210-320-6294
www.aquiferalliance.org/

San Marcos River Foundation

San Marcos River Foundation
222 W. San Antonio Street
San Marcos, TX 78666
512-353-4628

www.sanmarcosriver.org/

Save Our Springs Alliance

Save Our Springs Alliance
211 E. 9th Street, Suite 300
Austin, TX 78701
512-477-2320
www.sosalliance.org/

Texas Center for Policy Studies

Texas Center for Policy Studies
1002 West Avenue, Suite 300
Austin, TX 78707
512-740-4086
www.texascenter.org/

Texas Conservation Alliance

Texas Conservation Alliance
P.O. Box 6295
Tyler, TX 75711-6295
903-592-0909
www.tconr.org/

Texas Rivers Protection Association

Texas Rivers Protection
Association
444 Pecan Park Drive
San Marcos, TX 78666
512-392-6171
www.txrivers.org/

Texas Water Matters

800-919-9151
www.texaswatermatters.org/

Texas Wildlife Association

Texas Wildlife Association
2800 NE Loop 410, Suite 105
San Antonio, TX 78218
210-826-2904
www.texas-wildlife.org/

Wimberley Valley Watershed Association

Wimberley Valley Watershed
Association
P.O. Box 2534
Wimberley, TX 78676
512-847-1582

www.visitwimberley
.com/water/

American Rivers
American Rivers
1101 14th Street NW, Suite 1400
Washington, DC 20005
202-347-7550
www.americanrivers.org/

Coastal Conservation Association
Coastal Conservation
Association
6919 Portwest, Suite 100
Houston, TX 77024
713-626-4234
www.joincca.org/

Environmental Defense
Environmental Defense
44 East Avenue, Suite 304
Austin, TX 78701
512-478-5161
www.environmentaldefense
.org/

National Fish and Wildlife Foundation
National Fish and Wildlife
Foundation
1120 Connecticut Avenue NW,
Suite 900
Washington, DC 20036
202-857-0166
www.nfwf.org/

National Wildlife Federation
National Wildlife Federation
44 East Avenue, Suite 200
Austin, TX 78701
512-476-9805
www.nwf.org/

Sierra Club
Sierra Club
P.O. Box 1931
Austin, TX 78767
512-477-1729
http://texas.sierraclub.org

American Water Works

Association
American Water Works
Association
6666 W. Quincy Avenue
Denver, CO 80253
800-926-7337
www.awwa.org/

Association of Texas Soil and Water Conservation Districts
Association of Texas Soil and
Water Conservation Districts
311 N 5th Street
Temple, TX 76501
254-773-2250
www.tsswcb.state.tx.us/swcds

Texas Association of Flood Plain Managers
Mike Howard, CFM
Texas Water Development
Board
1700 North Congress Avenue
P.O. Box 13231
Austin, TX 78711-3231
512-463-3509
www.floods.org

Texas Alliance of Groundwater Districts
Texas Alliance of Groundwater
Districts
2104 Midway Court
League City, TX 77573
512-535-2126
www.texasgroundwater.org/

Texas Groundwater Association
Texas Groundwater
Association
221 E. 9th Street, Suite 206
San Jacinto Building
Austin, TX 78701
512-472-7437
www.tgwa.org/

Texas River and Reservoir

Management Society

Texas River and Reservoir
Management Society
8308 Elander Drive
Austin, TX 78750
512-347-1088
www.nalms.org/trrms/

Texas Rural Water Association

Texas Rural Water Association
1616 Rio Grande Street
Austin, TX 78701
512-472-8591
www.trwa.org/

Texas Water Conservation Association

Texas Water Conservation
Association
221 East 9th Street, Suite 206
Austin, TX 78701
512-472-7216
www.twca.org/

Texas Water Utilities Association

Texas Water Utilities
Association
1106 Clayton Lane,
Suite 112 West
Austin, TX 78723
512-459-3124
www.twua.org/

Water Environment Association of Texas

Water Environment Association of Texas
2619-C Jones Road
Austin, TX 78745
512-693-0060
http://www.weat.org/

Texas State Soil and Water Conservation Board

Texas State Soil and Water
Conservation Board
311 N. 5th Street
Temple, TX 76501-3107
254-773-2250
800-792-3485
www.tsswcb.state.tx.us/

BIBLIOGRAPHY

Adams, Jennifer, Dotti Crews, and Ronald Cummings. 2004. "The Sale and Leasing of Water Rights in Western States: An Update to Mid-2003." Georgia Water Policy and Planning Center. www.h2opolicycenter.org/pdf_documents/water_working papers/2004-004.pdf. (Accessed 2005.)

Addink, Sylvan. n.d. "Cash for Grass–A Cost Effective Method to Conserve Landscape Water." University of California at Riverside Turfgrass Research Facility. http://ucrturf.ucr.edu/topics/Cash-for-Grass.pdf. (Accessed 2005.)

American Heritage Dictionary. 2000. "Physiography." New York: Houghton Mifflin. www.bartleby.com/61/67/P0276700.html.

Angelina and Neches River Authority. n.d. Sam Rayburn Reservoir. www.anra.org/index_samrayburn.htm. (Accessed 2005.)

Blackburn, Jim. 2004. "Sustainability and Water for Texas." Unpublished speech delivered at the Global Environmental Forum on Water, Rice University, April 1, 2004.

———. 2004. *Texas Bays*. College Station: Texas A&M Press.

Blackburn, Jim, Mary Carter, and Francis Chin. n.d. "An Environmental Perspective on Texas Coastal Issues." www.clean houston.org/comments/editorials/jbb/coast.htm. (Accessed 2007.)

Boisseau, Charles. n.d. "The Saltcedar War." www.lcra.org/featurestory/2003/2003_5_21saltcedar.html. (Accessed 2005.)

Boyle, Robert H., John Graves, and T. H. Watkins. 1971. *The Water Hustlers*. San Francisco: Sierra Club.

Brune, Gunnar. 1981. *Springs of Texas*. Vol. 1. College Station: Texas A&M University Press

Carle, David. 2004. *Introduction to Water in California*. Berkeley: University of California Press.

Caroom, Douglas G. 1999. "Water Law in a Nutshell." www.bickerstaff.com/articles/waternut.htm. (Accessed 2003.)

Casteel, Pamela. 1994. "Conserving Corpus Christi Bay: Stewardhip Built by Consensus." www.cbbep.org/projectupdates/virtuallibrary/casteel/casteel.htm. (Accessed 2005.)

Chamberlin, Sean. n.d. "Water, Water, Everywhere: The Properties of Water." www.earthscape.org/tx/chs01/chs01e/chs01ea.html. (Accessed 2005.)

City of Beaumont. n.d. "Population Growth." www.bmtcoc.org/custom2.asp?pageid=1847

Coastal Bend Bays and Estuaries Program. 1998. *Coastal Bend Bays Plan*. Austin: Texas Natural Resource Conservation Commission.

Colorado Municipal Water District. n.d. "Water Quality Issues." www.crmwd.org/wqmain.htm. (Accessed 2005.)

Committee on Review of Methods for Establishing Instream Flows for Texas Rivers of the National Research Council of the National Academies. 2005. *The Science of Instream Flows: A Review of the Texas Instream Flow Program*. Washington, D.C: National Academies Press.

Congressional Budget Office. 1997. "Water Use Conflicts in the West: Implications of Reforming the Bureau of Reclamation's Water Supply." www.cbo.gov/showdoccfm?index=46&sequence=0. (Accessed 2005.)

Connecticut Department of Environmental Protection. 2005. "What Are PCB's?" http://dep.state.ct.us/wst/pcb/pcbindex.htm. (Accessed 2005.)

Dearen, Patrick. 1996. *Crossing the Pecos*. Fort Worth: Texas Christian University Press.

Detroit Free Press. 2005. "Granholm Action Gets Attention of Maine Water Campaign." www.freep.com/news/statewire/sw116856_20050607.htm. (Accessed 2005.)

Drennan, Todd. n.d. "Where the Action's At: The U.S.-Mexican Border." www.fas.usda.gov/info/agexporter/1999/articles/wherethe.html. (Accessed 2005.)

Eckhardt, Greg. n.d. "Desalination." www.edwardsaquifer.net/
desalination.html.

Edwards Aquifer Authority. 2005–2006. Various articles. http://
edwardsaquifer.org/. (Accessed 2005–2006.)

Edwards Aquifer Web Site. 2005–2006. Various articles. http://
www.edwardsaquifer.net/. (Accessed 2005–2006.)

El Paso Information and Links. n.d. "Population." www
.elpasoinfo.com/.

Encyclopaedia Britannica. 2007. "Sangre de Cristo Mountains."
www.britannica.com/eb/article-9065515/Sangre-de-
Cristo-Mountains.

Environmental Defense and Texas Center for Policy Studies.
2003–2004. Environmental profiles. Various topics. www
.texasep.org/. (Accessed 2003–2004.)

Environmental Institute of Houston. n.d. "State of the Bay: A
Characterization of the Galveston Bay Ecosystem." www.tsgc/
utexas.edu/topex/buoy/galveston.html. (Accessed 2005/2007.)

Estridge, Holli L. 2005. "Reuse Plans Could Hold the Line on
Costlier Options." *Dallas Business Journal.* www.bizjournals
.com/dallas/stories/2005/07/25/story8.html. (Accessed November 2006.)

Fisher, Lewis F. 1997. *Crown Jewel of Texas: The Story of the San
Antonio River.* San Antonio: Maverick Publishing Co.

Galveston Bay Information Center. n.d. "Frequently Asked Questions." http://gbic.tamug.edu/ss/faq.html. (Accessed 2005.)

Gamboa, Suzanne. n.d. "Sources Say Mexico, U.S. Reach Agreement on Water Debt." www.google.com/search?hl=en&q=r
etamal+diversion+dam+purpose+of+and+need. (Accessed
2005.)

Gerston, Jan, Mark MacLeod, and C. Allan Jones. 2002. "Efficient
Water Use for Texas: Policies, Tools, and Management Strategies." http://twri.tamu.edu/reports/2002/tr200/tr200.pdf. (Accessed 2005.)

Gilliland, Charles E., and Charles S. Middleton. 2004. "Westward
Ho: Recreational Buyers Explore New Territory." *Tierra
Grande: Journal of the Real Estate Center at Texas A&M University* 11, no. 2 (April 2004). www.recenter.tamu.edu/tgrande/
vol. (Accessed 2006.)

Gilliland, Charles E., John Robertson, and Heath Cover. 2004.
"Water Power." *Tierra Grande: Journal of the Real Estate Center at Texas A&M University* 11, no. 4 (October 2004). www.
recenter.tamu.edu/TGrande/vol11-4/1691.html. (Accessed
2006.)

Gonzalez, Martha. 2004. "The ABC's of Drought." www.txdpx
.state.tx.us/ftp/dem/digest/DIGESTFal2004.pdf. (Accessed
2005.)

Graves, John. 1960. *Goodbye to a River*. New York: Random
House.

Guadalupe-Blanco River Authority and Guadalupe River Trout
Unlimited. 2001. Contract between GRTU and GBRA. www
.grtu.org/GRTU_Signed_Contract.html. (Accessed 2006.)

Gulf of Mexico Foundation. n.d. "Facts about the Gulf of Mexico."
www.gulfmex.org/facts.htm. (Accessed 2005.)

Handbook of Texas Online. 2005–2006. Various articles. www.tsha
.utexas.edu/handbook/online/. (Accessed 2005–2007.)

Hess, Myron, Mary Kelly, and Ken Kramer. 2005. "Comments
on Initially Prepared 2006 Rio Grande Regional Water Plan."
www.texaswatermatters.org/pdfs/articles/2005_ipp_region_m_
joint_comments.pdf. (Accessed 2006.)

High Plains Water District. n.d. "The Ogallala Aquifer." www
.hpws.com/ogallala/ogallala/asp. (Accessed 2005.)

Hilson, Eileen Grevey. n.d. Testimony before the U.S. House of
Representatives Committee on Resources. http://resources
committee.house.gov/archives/108/testimony/eileenhillson.
htm. (Accessed 2005.)

Holloway, Milton. n.d. "Water Market Development in Texas: A
Prescription for Economic Efficiency." www.texaspolicy.com/
pdf/2004-PO-water-mh-PPT.pdf. (Accessed 2005.)

Horgan, Paul. 1954. *Great River: The Rio Grande in North American
History*. New York: Holt, Rinehart and Winston.

Hyer, Julien. 1952. *The Land of Beginning Again: The Romance of
the Brazos*. Atlanta: Tupper and Love.

International Boundary and Water Commission. 2002. "Report
of the United States Section." www.ibwc.state.gov/Files/
BrandesRpt0402.pdf.

———. 2003. 2003 Regional Assessment of Water Quality in the
Rio Grande Basin. http://www.ibwc.state.gov/CRP/2003_Rio_
Grande_Assessment.pdf.

———. n.d. "Environmental Assessment, Sediment Removal
Downstream of Retamal Diversion Dam." www.ibwc.state
.gov/EMD/FINAL_EA_Retamal/documents/section1.pdf.
(Accessed 2005.)

Jennings, Gregory D., Ronald E. Sneed, and May Beth St. Clair.
1996. "Metals in Drinking Water." www.bae.ncsu.edu/
programs/extension/publicat/wqwm/ag473_1.htm. (Accessed
2004.)

Johns, Dr. Norman. 2004. *Bays in Peril–A Forecast for Freshwater Flows to Texas Estuaries.* Washington, D.C.: National Wildlife Federation.

———. 2006. "The Potential and Promise of Municipal Water Effieiency Savings in Texas." Presentation at Texas Water Law Institute, December 7–8, 2006.

Kaiser, Ronald A. 1998. "A Primer on Texas Surface Water Law for the Regional Planning Process." www.bickerstaff.com/articles/waternut.htm. (Accessed 2003.)

LaCoast. n.d. "Calcasieu/Sabine Basin." lacoast.gov/geography/cs/index.asp. (Accessed 2005.)

Lesikar, Bruce, Ronald Kaiser, and Valeen Silvy. 2002. "Questions about Groundwater Conservation Districts in Texas." http://twri.tamu/reports/2002/2002-036-Questions-dist.pdf. (Accessed 2005.)

Longley, William L. 1994. *Freshwater Inflows to Texas Bays and Estuaries: Ecological Relationships and Methods for Determination of Needs.* Austin: Texas Water Development Board.

Lukens, Jennifer L. n.d. "National Coastal Program Dredging Policies." http://coastalmanagement.noaa.gov/resources/docs/finaldredge.pdf. (Accessed 2005.)

Mace, Robert E., Rima Petrossian, Robert Bradley, and William F. Mullican III. 2006. "A Streetcar Named Desired Future Conditions: The New Groundwater Availability for Texas." www.twdb.state.tx.us/GwRD/pdfdocs/03-1_mace.pdf. (Accessed 2005.)

MacLeod, Mark, and Elta Smith. n.d. "Economic Principles for Sound Water Planning–An Introduction for Regional Water Planning Groups in Texas." www.texaswatermatters.org/pdfs/economic_principles_report.pdf. (Accessed 2005.)

Martin, Quentin, Doyle Mosier, Jim Patek, and Cynthia Gorham-Test. 1997. *Freshwater Inflow Needs of the Matagorda Bay System.* Austin: Lower Colorado River Authority.

Mathis, Mitch. 2004. "How Do We Benefit from Inflows." Paper presented at the conference Water and the Future of the Texas Coast, September 25, 2004, Houston, Tex. http://texas.sierraclub.org/press/2004HoustonWaterConference.pdf. (Accessed 2005.)

McKinney, Larry. 2000. "Troubled Water." www.tpwd.state.tx.us/texaswater?sb1/primer/primer1/troubwat.phtml. (Accessed 2005.)

———. 2003. "Why Bays Matter." *Texas Parks and Wildlife Magazine,* July 2003, 23–25.

.aapg.org/explorer/2001/07jul/pickens_water.cfm. (Accessed 2005.)

Smith, Elizabeth H., Thomas R. Calnan, and Susan A. Cox. 1997. "Potential Sites for Wetland Restoration, Enhancement and Creation: Corpus Chrisi/Nueces Bay Area." Corpus Christi Bay National Estuary Program, CCBNEP-15.

Spearing, Darwin. 1991. *Roadside Geology of Texas*. Missoula: Mountain Press.

Stehn, Tom. 2001. "Relationship between Inflows, Crabs, Salinities, and Whooping Cranes." www.learner.org/jnorth/tm/crane/Stehn_CrabDocument.html. (Accessed 2005.)

Steiert, Jim. 1995. *Playas: Jewels of the Plains*. Lubbock: Texas Tech University Press.

Swanson, Eric R. 1995. *Geo-Texas: A Guide to the Earth Sciences*. College Station: Texas A&M University Press.

Texas Administrative Code. 2005–2007. Various articles. http://www.sos.state.tx.us/tac/ (Accessed 2005–2007).

Texas Center for Policy Studies. n.d. "The Gulf Intracoastal Waterway from Corpus Christi to Brownsville, Little Value, Big Cost." www.texascenter.org/publications/littlevaluebigcost.pdf. (Accessed 2005.)

Texas Coastal Wetlands. n.d. "Introduction to Coastal Wetlands." http://texaswetlands.org/introduction.htm. (Accessed 2005.)

Texas Commission on Environmental Quality. 2004. "Texas Nonpoint Source Pollution Management Program, SFR-066/03." http://www.tceq.state.tx.us. (Accessed 2005.)

———. 2004. "Get to Know the Arroyo Colorado." Texas Commission on Environmental Quality GI-318.

———. 2005–2007. Various articles. www.tceq.state.tx.us/. (Accessed 2005–2007.)

Texas Department of Transportation. n.d. "Gulf Intracoastal Waterway, 2003–2004 Legislative Report." www.dot.state.tx.us/publications/transportation_planning/giww03.pdf. (Accessed 2005.)

Texas General Land Office. 2005–2007. Various articles. www.glo.state.tx.us/. (Accessed 2005–2007.)

Texas House of Representatives, House Research Organization. 2006. "Groundwater Management Issues in Texas." www.hro.house.state.tx.us/focus/groundwater79-14.pdf. (Accessed 2006.)

Texas Legislature Online. Senate Bill 3. www.capitol.state.tx.us/billlookup/history.aspx?Leg Sess=80R+bill=SB3. (Accessed 2007.)

Texas Marine Advisory Service. n.d. "The Gulf Intracoastal Waterway–Texas' Unsung Hero in the Battle for Economic Independence." http://mts.tamug.tam.edu/GIWW/tti-giww.html. (Accessed 2005.)

Texas Natural Resource Conservation Commission. 2002. "Draft 2002 Water Quality Assessments for Individual Water Bodies." www.tnrcc.state.tx.us/water/quality/02_twpmar/02_305b/index.html. (Accessed 2005.)

———. 2004. *Draft 2004 Water Quality Inventory and 303d List.* Austin: Texas Environmental Conservation Commission.

Texas Nature Conservancy. 2001. "Conservation Plan for the Texas Portion of the Laguna Madre." http://conserveonline.org/docs/2002/02/Final_SCP_Laguna_Madre.doc. (Accessed 2005.)

Texas Parks and Wildlife Department. 1998. *Freshwater Inflow Recommendation for the Guadalupe Estuary.* Austin: Texas Parks and Wildlife.

———. 2001. *Freshwater Inflow Recommendation for the Trinity-San Jacinto Estuary.* Austin: Texas Parks and Wildlife.

———. 2002. *Freshwater Inflow Recommendation for the Nueces Estuary.* Austin: Texas Parks and Wildlife.

———. 2005. "Fish Consumption Bans and Advisories." www.tpwd.state.tx.us/publications/annual/fish/bans-advisories.phtml. (Accessed 2005.)

———. 2005. "Golden Alga FAQ." www.tpwd.state.tx.us/hab/ga/faq.phtml. (Accessed 2005.)

———. 2005–2006. Various articles. www.tpwd.state.tx.us/. (Accessed 2005–2006.)

———. n.d. "Environmental Target Flows." www.tpwd.state.tx.us/texaswater/sb1?enviro/envwaterneeds/envwaternees.html. (Accessed 2005.)

———. n.d. "Texas Coastal Treasures–Vol I." www.tpwd.state.tx.us/fish/recreat/faq/tct.htm. (Accessed 2005.)

Texas State Water Plan. 2007. "Chapter 6: Surface Water Resources." www.twdb.state.tx.us/publications/reports/State_Water_Plan/2007/2007%20final%20draft%20SWP/Chapter%206_FINAL%20110706.pdf. (Accessed 2007.)

Texas Water Code. 2005–2007. Various articles. http://tlo2.tlc.state.tx.us/statutes/wa.toc.htm. (Accessed 2005–2007.)

Texas Water Development Board. 2001. "2001 Water Use Survey Summary Estimates for Region C in acft." www.twdb.state.tx.us/data/popwaterdemand/2003Projections/HistoricalWaterUse/2001WaterUse/HTML/2001RegionC.htm.

———. 2005. "Water for Texas 2002." www.twdb.state.tx.us/ publications/reports/State_Water_Plan/2002/FinalWaterPlan 2002.asp. (Accessed 2005–2006.)

———. 2006. "Texas Water Trust." www.twdb.state.tx.us/ assistance/WaterBank/wtrust.asp.

———. 2005–2006. Various articles. www.twdb.state.tx.us/home/ index.asp. (Accessed 2005–2006.)

———. 2006. "Water for Texas 2007." www.twdb.state.tx.us/ publications/reports/State_Water_Plan/2007/2007 StateWaterPlan/2007StateWaterPlan.htm. (Accessed 2007.)

———. 2006. "Springs of Kinney and Val Verde Counties: Report to Region J Water Planning Group." www.twdb.state.tx.us/ rwpg/2006_RWP/RegionJ/Kinney%20Spring%20Study/ Kinney_Spring_Report_Text.doc. (Accessed 2006.)

———. 2006. "2006 Adopted Regional Water Plans." www.twdb .state.tx.us/rwpg/main-docs/2006RWPindex.asp. (Accessed 2006.)

Texas Water Foundation. 2005. Various articles. www.texaswater .org. (Accessed 2005.)

Texas Water Matters. 2005–2007. Various articles. www.texas watermatters.org/. (Accessed 2005–2007.)

Texas Water Safari. 2004. Various articles. www.texaswater safari.org/.

TopoZone. 1999–2006. Various maps. www.topozone.com/map .asp?lon=-97.0892&lat=27.8731.

Trinity River Authority. n.d. "TRA Makes First Payment on Wallisville Saltwater Barrier Project." www.trinityra.org/PDF_ files/News_Dec%202003-Jan%202004%20Newsletter.pdf. (Accessed 2005.)

Union of Concerned Scientists. n.d. "Laguna Madre, Salty Balance." www.ucsus.org/gulf/gcplaceslag.html. (Accessed 2005.)

United States Army Corps of Engineers. 2005–2006. Various articles. www.usace.army.mil/. (Accessed 2005–2006.)

———. n.d. Houston-Galveston Navigation Channel Project Online Resource Center. www.swg.usace.army.mil/items/hgnc/. (Accessed 2005.)

———. n.d. "A Plan for National Transportation." www.usace .army.mil/publications/misc/nws83-9/c-1.pdf. (Accessed 2005.)

United States Environmental Protection Agency. 2005. "Clean Air Mercury Rule–Basic Information." www.epa.gov/air/ mercuryrule/basic.htm. (Accessed 2005.)

———. 2005–2007. Various articles. www.epa.gov/. (Accessed 2005–2007.)

———. n.d. "Galveston Bay Estuary Program." www.epa.gov/owow/estuaries/programs/gb.htm. (Accessed 2005.)

———. n.d. "Rio Grande: The Flow of the Mighty River Down to the Gulf of Mexico." www.epa.gov/region6/water/riogrande/brochure.pdf. (Accessed 2005.)

———. n.d. "Fecal Coliform." www.epa/gov/maia/html/fecal.html. (Accessed 2005.)

United States Fish and Wildlife Service. 1996. "History and Evolution of the Endangered Species Act of 1973." http://endangered.fws.gov/esasum.html. (Accessed 2005.)

———. 2001. "2001 National Survey of Fishing, Hunting, and Wildlife-Associated Recreation–Texas." www.census.gov/prod/2003pubs/01fhw/fhw01-tx.pdf.

———. 2005–2006. Various articles. www.fws.gov/. (Accessed 2005–2006.)

United States Geological Survey. 2005–2007. Various articles. www.usgs.gov/. (Accessed 2005–2007.)

———. n.d. "Edwards-Trinity Aquifer System." www.capp.water.usgs.gov/gwa/ch_e/E-text8.html. (Accessed 2005.)

———. n.d. "Environmental Change in South Texas." http://biology.usgs.gov/s+t/SNT/noframe/se132.htm. (Accessed 2005.)

University of Texas at Austin. 2001. "Water World–A Trip to UT's Marine Science Institute at Port Aransas and Beyond." http://utopia.utexas.edu/articles/alcalde/water_world.html?sed=texas&sub+university. (Accessed 2005.)

Votteler, Todd H. 1998. "The Little Fish That Roared: The Endangered Species Act, State Groundwater Law, and the Private Property Rights Collide over the Edwards Aquifer." www.gbra.org/Documents/EA/Policy/LittleFish.pdf. (Accessed 2006.)

———. n.d. "Flood." www.gbra.org/documents/Publications/Flood.pdf. (Accessed 2005.)

———. n.d. "Texas Water." www.texaswater.org/water/drought/drought2.htm. (Accessed 2005.)

Walter Geology Library. n.d. "Physiography of Texas." www.lib.utexas.edu/geo/physiography.html. (Accessed 2005.)

Wassenich, Tom. 2004. "The State of the Protection of Freshwater Inflow to the Bays and Estuaries of Texas, 2003." M.A. thesis, Texas State University–San Marcos.

Water Conservation and Implementation Task Force. 2004. "Water Conservation Implementation Task Force Report to the 79th Legislature." www.twdb.state.tx.us/assistance/conservation/TaskForceDocs/WCITF_Leg_Report.pdf. (Accessed 2006.)

————. 2004. "Water Conservation Best Management Practices Guide." www.twdb.state.tx.us/assistance/conservation/TaskForceDocs/WCITFBMPGuide.pdf. (Accessed 2006.)

Wilson, Robert H. 1999. "Understanding Urban Texas." www.utexas.edu/opa/pubs/discovery/disc1998v15n2/disc_urban.html. (Accessed 2005.)

Winemiller, Kirk O., Soner Tarim, James B. Cotner, and David E. Shormann. n.d. "Brazos Oxbow Community Structure." www.sdafs.org/tcafs/meetings/96meet/winemill.htm. (Accessed 2005.)

Younos, Tamim. 2005. "Environmental Issues of Desalination." *Journal of Contemporary Water Research and Education* 132 (December 2005). www.ucowr.siu.edu/updates/132/3.pdf. (Accessed 2006.)

INDEX